Data-Driven Solutions to Transportation Problems

Data-Driven Solutions to Transportation Problems

Edited by

Yinhai Wang
University of Washington

Ziqiang Zeng
Sichuan University

ELSEVIER

Elsevier
Radarweg 29, PO Box 211, 1000 AE Amsterdam, Netherlands
The Boulevard, Langford Lane, Kidlington, Oxford OX5 1GB, United Kingdom
50 Hampshire Street, 5th Floor, Cambridge, MA 02139, United States

Library of Congress Cataloging-in-Publication Data
A catalog record for this book is available from the Library of Congress

British Library Cataloguing-in-Publication Data
A catalogue record for this book is available from the British Library

ISBN: 978-0-12-817026-7

Publisher: Joe Hayton
Acquisition Editor: Biran Romer
Editorial Project Manager: Ali Afzal-Khan
Production Project Manager: Anitha Sivaraj
Designer: Christian J. Bilbow

Typeset by SPi Global, India

Contents

Contributors

Numbers in Parentheses indicate the pages on which the author's contributions begin.

Matthew J. Barth (11), Department of Electrical and Computer Engineering; College of Engineering-Centre for Environmental Research and Technology (CE-CERT), University of California, Riverside, CA, United States

Kanok Boriboonsomsin (11), College of Engineering-Centre for Environmental Research and Technology (CE-CERT), University of California, Riverside, CA, United States

Xi Chen (175), School of Transportation Science and Engineering, Beihang University, Beijing, People's Republic of China

Xiqun (Michael) Chen (201), College of Civil Engineering and Architecture, Zhejiang University, Hangzhou, People's Republic of China

Ge Guo (247), Institute of Computing Technology, China Academy of Railway Sciences, Beijing, People's Republic of China; Department of Civil and Environmental Engineering, University of Washington, Seattle, WA, United States

Meng Li (111), Department of Civil Engineering, Tsinghua University, Beijing, People's Republic of China

Huiping Li (111), Department of Civil Engineering, Tsinghua University, Beijing, People's Republic of China

Li Li (247), Institute of Computing Technology, China Academy of Railway Sciences, Beijing, People's Republic of China

Xiaolei Ma (175), School of Transportation Science and Engineering, Beihang University, Beijing, People's Republic of China

Xuewei Qi (11), Department of Electrical and Computer Engineering; College of Engineering-Centre for Environmental Research and Technology (CE-CERT), University of California, Riverside, CA, United States

Haiyan Shen (247), Institute of Computing Technology, China Academy of Railway Sciences, Beijing, People's Republic of China

Tianyun Shi (247), Institute of Computing Technology, China Academy of Railway Sciences, Beijing, People's Republic of China

Xiaoqian Sun (227), National Key Laboratory of CNS/ATM, School of Electronic and Information Engineering, Beihang University, Beijing, People's Republic of China

Peng Sun (247), Institute of Computing Technology, China Academy of Railway Sciences, Beijing, People's Republic of China

Jinjun Tang (137), School of Traffic & Transportation Engineering, Central South University, Changsha, China

Sebastian Wandelt (227), National Key Laboratory of CNS/ATM, School of Electronic and Information Engineering, Beihang University, Beijing, People's Republic of China

Yinhai Wang (1,51), Department of Civil and Environmental Engineering, University of Washington, Seattle, WA, United States

Guoyuan Wu (11), College of Engineering-Centre for Environmental Research and Technology (CE-CERT), University of California, Riverside, CA, United States

Yao-Jan Wu (81), Department of Civil and Architectural Engineering and Mechanics, University of Arizona, Tucson, AZ, United States

Shu Yang (81), Department of Civil and Architectural Engineering and Mechanics, University of Arizona, Tucson, AZ, United States

Ziqiang Zeng (1), Department of Civil and Environmental Engineering, University of Washington, Seattle, WA, United States, Business School, Sichuan University, Chengdu, People's Republic of China

Guohui Zhang (51), Department of Civil and Environmental Engineering, University of Hawaii at Manoa, Honolulu, HI, United States

Mingqiao Zou (111), Department of Civil Engineering, Tsinghua University, Beijing, People's Republic of China

List of Figures

List of Tables

Preface

In recent years, the increasing quantity and variety of data available for decision support present a wealth of opportunity as well as a number of new challenges, in both the public and private sectors. Vast quantities of data are available through increasingly affordable and accessible data acquisition and communication technologies, including sensors, cameras, mobile location services, etc. When these are combined with emerging computing and analytical methodologies, they can lead to more thorough scientific understandings, informed decisions, and proactive management solutions. As a result, big data concepts and methodologies are steadily moving into the mainstream in a variety of science and engineering fields.

During the past decades, transportation research has been driven largely by mathematical equations and has relied on relatively scarce data. With the increasing quantity and variety of data being collected from intelligent transportation systems and other sensors and applications, the potential for solid data-driven or data-based research is increasing rapidly. Nevertheless, today there are few established systems for supporting general big data analytics in transportation research and practical applications. Most current online data analysis and visualization systems are designed to compute and visualize one type of data, such as those from freeway or arterial sensors, on an online platform. Therefore, though the scope and ubiquity of transportation data are increasing, making these data accessible, integrated, and useable for transportation analysis is still a remarkable challenge.

Understanding data-driven transportation science is essential for enhancing an intelligent transportation system's performance. Most commercial systems are oriented toward a specific transportation problem or analysis procedure, and approach the problem in their own (often ad hoc) way. A mature framework for effectively utilizing data and computing resources, such that these data will serve the needs of users, has become a pressing need in the field of transportation. The challenges associated with developing this type of framework primarily stem from the need for standardized and efficient data integration and quality control methods, computational modules for applying these data to transportation analysis, and a unified data schema for heterogeneous data.

This book consists of 10 chapters providing in-depth coverage of the state of the art in data-driven methodologies and their applications in the *E*-Science of transportation. Such methods are crucial for solving transportation problems such as energy-efficient driving in a connected vehicle environment, traffic sensing data analysis and quality enhancement, travel time reliability (TTR) estimation, urban travel behavior and mobility analysis, public transportation data mining, network modeling, and railway system prognostics and health management (PHM).

A brief overview of chapters in this book is provided here as a quick guide for readers. The structure and connections between different chapters are also illustrated in a roadmap to help the readers gain a better understanding of the content of this book.

Chapter 1 presents an overview of data-driven transportation science. A general background on the motivation for promoting data-driven transportation science is provided. In addition, a review of related methodologies and applications is given as an introduction to the development history of intelligent transportation systems.

Chapter 2 introduces two data-driven on-line energy management strategies for plug-in hybrid electric vehicles (PHEVs), which support energy-efficient driving control in a connected vehicle environment. The methods introduced in this chapter are validated using real-world driving data, and the results indicate that the proposed data-driven energy management system (EMS) strategies are very promising in terms of achieving a good balance between real-time performance and fuel savings when compared with some existing strategies, such as binary mode EMS and Dynamic Programming-based EMS.

Chapter 3 describes an artificial neural network-based machine learning method to extract classified vehicle volumes from single-loop measurements. In addition, a set of computer vision-based algorithms is developed to extract background images from a video sequence, detect the presence of vehicles, identify and remove shadows, and calculate pixel-based vehicle lengths for classification based on widely available surveillance camera signals. Machine learning methods for predictive modeling and computer vision are advanced computing techniques, which can revolutionize existing traffic sensing practices and theoretical foundations. The experimental results described in this chapter indicate that such methods exhibit superior performance under various traffic operation scenarios. This chapter summarizes current efforts in these promising areas, and offers significant contributions to data-driven transportation science research and applications.

Chapter 4 empirically demonstrates the concept that "the same data tell you the same story," and that TTR measures are insensitive to probability distribution assumptions. This chapter also covers accuracy estimation for TTR measures. The bootstrap technique, a data-driven technique based on resampling with replacement, plays an important role in accuracy estimation. The accuracy estimates provide a more general characterization of TTR compared to point estimation. In addition, the concept of segment-based TTR on roadways is extended to Origin-Destination (OD)-based TTR over roadway networks. The characteristics of OD-based TTR are discussed briefly. This chapter summarizes continued efforts on improving the accuracy of TTR estimation and related extensions, contributing to data-driven transportation studies and applications.

Chapter 5 covers some conventional methods for modeling travel behavior, and introduces several state-of-the-art analytical methods to study travelers' behaviors based on a data fusion method. Some traditional behavior models are based on the max-utility theory and perfect human rationality. The most widely used travel behavior model based on the maximization theory is the discrete choice model. This is operationalized in the modeling structure by making the choice process a function of both the alternative attributes and the characteristics of the traveler. Furthermore, analytical travel behavior models are used to predict travelers' departure time choice and mode switch under such strategies. Agent-based models for traveler mode choice and departure time are utilized in this chapter.

Chapter 6 explores the urban travel mobility for understanding the property of travel patterns based on large-scale trajectory data. By dividing the city area into different transportation districts, the origin and destination distribution associated to these districts in an urban area on weekdays and weekends are analyzed. The Density-Based Spatial Clustering of Applications with Noise (DBSCAN) algorithm is used to cluster pick-up and drop-off locations. Furthermore, four spatial interaction models are calibrated and compared based on trajectories in a shopping center of Harbin city to study the pick-up location searching behavior. By extracting taxi trips from GPS data, travel distance, time, and average speed in occupied and nonoccupied vehicles are then used to investigate human mobility. Next, the observed OD matrix of a central area in Harbin city is used to model the traffic distribution patterns based on the entropy-maximizing method and to validate the performance of the proposed methodology in a case study. Finally, a dilatation index based on the weighted average distance among trips is applied to analyze the spatial structure of an urban city. Furthermore, hotspots are identified from local density of locations with different thresholds as determined by the Lorenz curve.

In Chapter 7, applications of big data in public transportation planning, operation, and management is introduced, specifically with regard to the classification and processing of these big data and their combination with other data. Applications of public transportation big data in areas such as bus arrival times prediction, commuting behavior mining, and performance evaluation of public transportation networks (*E*-Science public transportation big data platform) are introduced. In addition, case studies are presented to demonstrate the value of Beijing's public transportation data in addressing practical problems.

Chapter 8 develops a simulation-based optimization (SBO) framework by integrating metamodels with mesoscopic simulation-based dynamic traffic assignment models for large-scale network modeling problems. The adopted SBO approach reconstructs the response surface by only a few evaluations of the objective function and is capable of handling simulation noises. This approach can result in computational timesavings, which are achieved through the use of metamodels to construct response surfaces for predicting optimal solutions. This chapter provides a macroscopic understanding of urban traffic dynamics using both a simulation-based dynamic traffic assignment model and heterogeneous traffic detection data. The simulation is validated by a representation of macroscopic fundamental diagrams using fixed traffic flow detections and probe travel time measurements. The SBO approach is demonstrated in a real-world large-scale transportation network that consists of arterials and freeways.

Chapter 9 describes the design, implementation, and dissemination of an open-source framework for analyzing the performance and resilience of air transportation networks. First, a framework for modeling air transportation networks based on freely available datasets is derived. Second, an overview on estimating the resilience of such a complex system is provided, with methods developed in the network science community. Third, experiments on global air transportation are performed, reporting on critical roles of its elements. The proposed framework, implemented in Python, makes it easy for transportation researchers to get started in the area of air transportation network resilience, by having a gold standard as a reference. Moreover, since the framework and its underlying data are freely available, this can push the state of the art in air transportation network resilience analysis.

Chapter 10 implements the railway system electric multiple units (EMU) health assessment from the data point of view using data fusion technology. As one of the most important types of passenger transport equipment, EMU's safety insurance is vital and the use of PHM technology is a suitable method. Because of the high speed, high geographical span, complicated operating environment, and long continuous running time, it is difficult to consider the influencing factors comprehensively when analyzing failure mechanism and build model to assess the health status of EMU. EMU's on-board monitoring system is relatively mature; hundreds of sensors collect various data continuously while EMU is running, and a huge amount of data has been accumulated, which can support data-driven health assessment.

In summary, this book showcases recent innovative attempts in applying data-driven methods to important problems in different transportation modes. Methodologies employed in these studies include data fusion, data mining, machine learning, etc. Readers may get hints on how data-driven methodologies have been applied in transportation research and practice. Researchers, practitioners, graduate students, and upper-level undergraduates with backgrounds in transportation engineering, management science, operations research, and engineering management may benefit from reading this book.

Yinhai Wang
Ziqiang Zeng
University of Washington

Acronyms

AAT	actual arrive time
ABM	agent-based modeling
ADP	approximate dynamic programming
AFC	automatic fare collection
AGC	automatic gain control
AIC	Akaike information criterion
ANN	artificial neural network
AVL	automated vehicle location
BI	buffer index
BIC	Bayesian information criterion
DBSCAN	density-based spatial clustering of applications with noise
DfT	Department for Transport
DOT	Department of Transportation
DOW	day of the week
DP	dynamic programming
EA	evolutionary algorithm
EBM	equation-based modeling
ECU	electronic control unit
EDA	estimation distribution algorithm
EMS	energy management system
FHWA	Federal Highway Administration
FTP	file transfer protocol
GIS	geographic information system
GMT	Greenwich Mean Time
HEVs	hybrid electric vehicles
IAA	irrelevant alternatives
ICE	internal combustion engine
ILD	inductive loops detector
ISODATA	iterative self-organizing data analysis technique algorithm
ITS	intelligent transportation systems
JPEG	joint photographic experts group
KDE	kernel density estimation
LHS	Latin Hypercube Sampling
LVs	long vehicles

MARs	multiple airport regions
MFD	Macroscopic Fundamental Diagram
MOVES	MOtor Vehicle Emission Simulator
MOY	month of year
NL	nested logit
NRS	non-route-specific
NSF	National Science Foundation
OBT	outside bus time
OD	origin-destination
OMT	outside metro time
OTTN	occupied trips based travel network
PAT	preferred arrival time
PeMS	performance measurement system
PHEV	plug-in hybrid electric vehicle
PHM	prognostics and health management
PM	particulate matters
RBF	radial basis function
RL	reinforcement learning
RP	revealed-preference
RVM	relevance vector machine
SBO	simulation-based optimization
SIM	subscriber identity module
SOC	state-of-charge
SVs	short vehicles
TD	temporal-difference
TOD	time of day
TOPSIS	technique for order of preference by similarity to ideal solution
TSB	technology strategy board
TTR	travel time reliability
VIPs	video image processors
VOS	visualization of similarities
VTTN	vacant trips based travel network
VVDC	video-based vehicle detection and classification
WSDOT	Washington State Department of Transportation

Chapter 1

Overview of Data-Driven Solutions

Yinhai Wang* **and Ziqiang Zeng***,†
*Department of Civil and Environmental Engineering, University of Washington, Seattle, WA, United States, †Business School, Sichuan University, Chengdu, People's Republic of China

Chapter Outline

1.1 GENERAL BACKGROUND

Data is essential to the planning, delivery, and management of issues related to transportation mobility, safety, and environment [1]. Nowadays, instead of relying on conventional mathematical models and traffic theory based on relatively scant data, transportation research is increasingly data-driven. Advances in sensors, telecommunications, and connected vehicles are making vast new data resources accessible to transportation researchers and practitioners. With the growing quantity and variety of data being collected from intelligent transportation systems (ITS) and other technologies, data-driven transportation research must rely on a new generation of tools to analyze and visualize those data. If all of these data can be brought together in a unified, dynamic, and real-time flow of information, it will revolutionize traveler decision-making and operations management.

This emerging trend will drive significant changes, not only in the methods of transportation research, but also in our way of thinking about and fundamental understanding of transportation systems. In this book, we define this trend as "data-driven transportation science." It should be noted that transportation

Data-Driven Solutions to Transportation Problems. https://doi.org/10.1016/B978-0-12-817026-7.00001-1

science has a very wide definition. The basic definition of transportation science is to make a transportation analysis by looking at all levels of decision-making in planning. These are analytical-, operational-, tactical-, and strategic-level transportation planning. The scope of this book will focus mostly on analytical-, tactical-, and operational-level planning. In fact, the development and improvement of our transportation systems follows two paths: a "hard path" that consists primarily of infrastructure design and construction with related hardware technology development, and a "soft path" that complements the former by investing in efficient traffic control, network optimization, and transport policies. While we believe that data-driven transportation science offers substantial opportunities in both paths, this book will focus mainly on the impacts on the soft path. Actually, governments, the academic community, and the transportation industry have been moving quickly to address the challenges associated with moving toward a data-driven transportation era. For the major investments that will be needed to facilitate this shift, decision-makers must turn to the wealth of data available and let it guide decisions as we build the transportation systems that will carry us into the next century. In the following subsections, we highlight some key examples of data-driven transportation decisions from a variety of focus areas.

1.1.1 Government Investment

Agencies and researchers around the world are focusing more attention on data-driven transportation. The United States (US) government spent approximately $128.4 billion on transportation in 2014. In 2016, the US Department of Transportation (DOT) selected Columbus, Ohio to receive $40 million to prototype the future of urban transportation, out of 78 cities participating in its Smart City Challenge. The city's plan, which will also leverage over $100 million in private resources, involves piloting a variety of new technologies. Such technologies include connected vehicles that improve traffic flow and safety, data-driven efforts to improve public transportation access and health care outcomes, and electric self-driving shuttles that will create new transportation options for underserved neighborhoods [2].

Also in 2016, the Chinese government collaborated with the transportation-related industry and data companies to establish a cloud-based big data transportation platform. China's internet giant Baidu Inc. launched an open platform dedicated to building an intelligent transportation cloud ecology including aviation, railway, and highway [3].

In the United Kingdom (UK), to maximize these opportunities, the government has supported the UK's data infrastructure since 2014 in order to leverage opportunities in data-driven decision-making. Most recently, this program invested £14 million to make data routinely collected by business and local government accessible for researchers, including for transportation research at Leeds and Glasgow Universities. The government has also established a new

Transport Systems Catapult, overseen by the Technology Strategy Board (TSB). This program has specific objectives to encourage the analysis of big data [4], and over 5 years will receive £46.6 million from TSB and £16.9 million from the Department for Transport (DfT). These data-driven improvements to transportation are not just about convenience; they also have a significant impact on economic potential and competitiveness [5].

1.1.2 Academic Community Research Trend

In the USA, the National Science Foundation (NSF) invested over $60 million in new smart cities-related grants in FY16 and planned new investments in FY17, in which big data research for transportation is a prioritized area [2]. Zhang et al. [6] conducted a survey on research for data-driven ITS, and summarized the research trends in different categories. Their results indicated that while vision and learning-driven ITS have received much attention from researchers in the ITS community, there is still room for further research directly addressing issues in data-driven ITS, such as multimodal evaluation criteria, visual analytics, and microblogs.

1.1.3 Transportation Industry Involvement

Transportation deficiencies impact all industries and citizens. Beyond impacts on the private sector, investments in data-driven transportation systems are needed to address the geographic population shift occurring as more and more people move from rural to urban areas. The latest census data shows that nearly 81% of all Americans live in cities and suburbs. This ongoing movement of people demands transportation systems capable of handling and moving a growing number of people [5].

Many companies operating in the transportation industry are focusing on data-driven transportation. Take the example of Bridj, a data-driven bus line tested in Massachusetts in the cities of Brookline, Boston, and Cambridge. The company seeks to offer a "pop-up" bus system that is tailored to where people work and live, and can rapidly adapt to changing demand. Using the wealth of data online, as well as consumer input, Bridj predicts areas of peak demand and adjusts bus service to satisfy it [5].

Just as with many other industries, railroad companies have integrated big data into many different aspects of their operations. As an example of railway automation, one of the nation's largest railroads just invested in a fully automated rescheduling system. This big data system manages the rescheduling of over 8000 trains to insure on-time operation across 23 states under a variety of planned and unplanned scenarios [7].

Freight delivery and trucking companies also have implemented big data technologies in order to keep up with the high expectations of their customers. One of the ways in which big data is reducing costs in the trucking industry is with fuel consumption. In some cases, mathematical models are used to optimize shipping routes. By focusing on excessive driving routes, drivers can see a reduction of nearly 1 mile of driving every day. This may not seem like much; however, for a company like UPS, a reduction of 1 mile per day per driver would equal savings of as much as $50 million a year in fuel [7].

Big data has helped transportation companies stay on track through increased operational efficiency, improved customer experiences, reduced fuel costs/increased profits, and enhanced service offerings [7].

1.2 DATA-DRIVEN INNOVATION IN TRANSPORTATION SCIENCE

Data-driven innovation entails exploitation of any kind of data in the innovation process to create value [8]. Emerging computing technology and analytical methods give us the ability to monitor traffic networks with greater coverage and granularity, and promise to improve the accuracy of traffic prediction [9].

In transportation systems, the number of data sources is increasing rapidly [10]. Take the City of Dublin as an example. The city's road and traffic department is able to combine big data streaming from an array of sources—including bus timetables, inductive loop traffic detectors, closed-circuit television cameras, and GPS updates that each of the city's 1000 buses transmits every 20 s—to build a digital map of the city overlaid with the real-time positions of Dublin's buses using stream computing and geospatial data. Some interventions have led to a 10%–15% reduction in journey times [11].

Data-driven innovation in transportation science follows two primary approaches: technology-oriented and the methodology-oriented (see Fig. 1.1). The technology-oriented approach focuses mainly on developing new sensor, communication, detection, and connected and autonomous vehicle related technologies. Typical examples include autonomous data driven surveillance and rectification system by using artificial intelligence-based techniques [12] and artificial intelligence for managing electric vehicles in the smart grid [13]. The methodology-oriented approach concentrates mostly on studying new analytical methods to get insights from the big data collected from the transportation system. Typical examples include deep-learning architecture to forecast destinations of bus passengers [14] and a deep learning-based rear-end collision prediction scheme [15]. Recently, many innovators have been trying to combine the two approaches by developing integrated data-driven transportation decision support platforms. They use the technology-oriented approach to enhance the data resources available to the platform, and employ the methodology-oriented

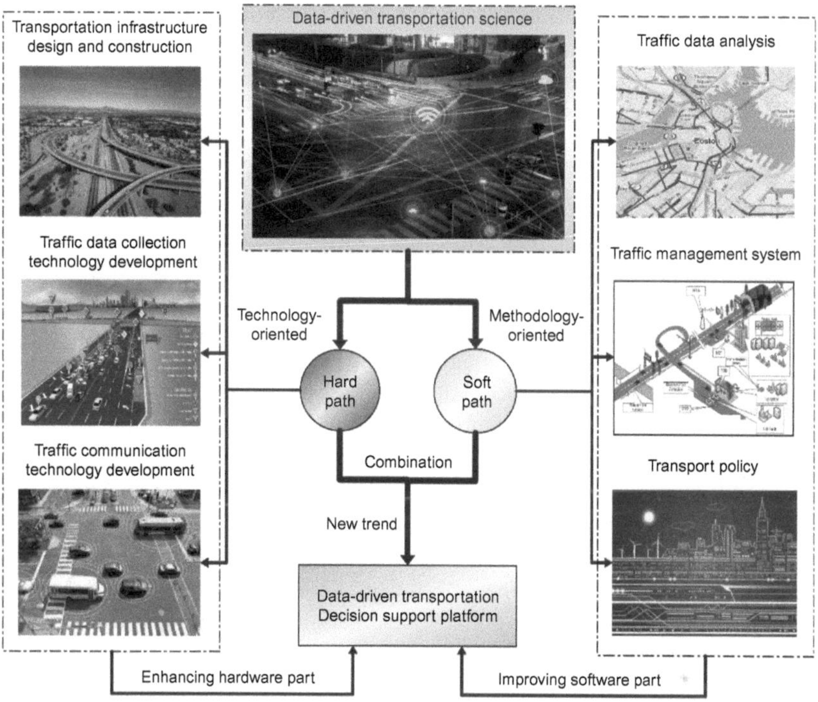

FIG. 1.1 Data-driven innovation process in transportation systems.

approach to improve the software part of the platform. This combined innovation can create great value and will likely grow in importance in the coming years.

1.3 METHODOLOGIES FOR DATA-DRIVEN TRANSPORTATION SCIENCE

Many data-driven methodologies have been developed and employed for addressing problems in transportation science. Chowdhury et al. summarized the state of the art in data analytics methods for ITS [16]. In their book, data science tools, data analytics approaches, and machine learning are introduced and discussed for ITS applications. Due to the rapid development of knowledge in this area, it is quite difficult to summarize all the important methodologies within one book; thus, this book will introduce the latest frontier of the data-driven transportation science as an update of the research area.

With the increasing size and complexity of traffic data from various sources, -data-learning-based models have drawn increasing attention from transportation researchers due to their ability to extract insightful information from the data

[17]. Different from traditional physical models that attempt to build mathematical structures based on causality, data-learning methods aim to establish the correlations between the inputs and outputs from field data. The principle of data-learning models is the correlations in the data, which refers to any of a broad class of statistical relationships involving dependence. These focus on explaining and representing the system by the data itself. The knowledge and the data are involved at the beginning of the modeling process. Normally, a highly representative basis function is established and trained with the data to extract statistically significant information fully. The domain knowledge is not specified through the mathematical structure. Instead, the empirical features are normally injected into the model by imposing certain constraints. Ghofrani et al. [18] summarized the recent models of big data analytics applied in railway transportation systems, including association models [19], clustering models [20], classification models [21], pattern recognition models [22], time series [23], stochastic models [24], optimization-based methods [25], and so on. Big data analytics has increasingly attracted a strong attention of analysts, researchers, and practitioners in transportation engineering.

This book summarized several useful data-driven methodologies that focus on addressing problems such as energy efficient driving control, traffic sensor data analysis, travel time reliability (TTR) estimation, urban travel behavior and mobility study, public transportation, gating control, and network modeling.

1.4 APPLICATIONS IN DATA-DRIVEN TRANSPORTATION SCIENCE

The summary provided in Rusitschka and Curry [11] suggests that big data applications in transportation systems can be categorized as operational efficiency, customer experience, and new business models, where operational efficiency is the main driver behind the investments for data-driven transportation science [26]. Ma and Wang [27] developed a data-driven platform for transit performance measures using smart card and GPS data. Tak et al. [28] developed a data-driven framework for real-time travel time prediction. Perugu et al. [29] employed integrated data-driven modeling to estimate PM2.5 pollution from heavy-duty truck transportation activity over a metropolitan area. Woo et al. [30] developed a data-driven prediction methodology for origin-destination demand in a large network for a real-time transportation service. Khadilkar [31] employed data-enabled stochastic modeling for evaluating the schedule robustness of railway networks. Haider et al. [32] used a data-driven method to develop the inventory rebalancing through pricing in public bike-sharing systems.

From a transportation systems perspective, most of the data-driven methodologies are applied in the following areas: transportation management systems, traveler information analysis, vehicle control and management, public transportation systems optimization, and urban transportation systems optimization.

From a data science perspective, these methodologies are mainly used to address problems such as data cleansing and imputing, data fusion, and heterogeneous data analysis.

1.5 OVERVIEW AND ROADMAP

The topics described in this book can be connected to two perspectives: data-driven methodologies and the applications. Each of the chapters will focus on the two perspectives to tell a compelling story. In Chapter 2, two data-driven on-line energy management strategies for plug-in hybrid electric vehicle (PHEV) energy-efficient driving control in a connected vehicle environment are introduced. Chapter 3 describes a machine learning approach to establish an artificial neural network to extract classified vehicle volumes from single-loop measurements more efficiently. Chapter 4 empirically demonstrates the concept that "the same data tells you the same story," and that TTR measures are insensitive to the probability distribution selection. Chapter 5 covers some of the typical approaches to modeling travel behavior, and introduces several state-of-the-art analytical methods to study travelers' behaviors based on a data fusing method. Chapter 6 analyzes the origin and destination distribution in urban area on weekdays and weekends by dividing the city area into different transportation districts. In Chapter 7, we introduce the application of big data in public transportation planning, operation, and management, as well as the classification and processing of these big data and their combination with other data. Chapter 8 develops a simulation-based optimization (SBO) framework by integrating metamodels with mesoscopic simulation-based dynamic traffic assignment models for large-scale network modeling problems. Chapter 9 designs, implements, and disseminates an open-source framework for the analysis of air transportation networks, their performance, and their resilience. Chapter 10 implements the railway system EMU health assessment from the data point of view using data fusion technology. Fig. 1.2 shows a roadmap guiding the readers to provide a better understanding of the structure of this book. Five data-driven methodologies are introduced including data-driven control and optimization (Chapters 2 and 9), data-driven learning (Chapter 3), data-driven estimation (Chapters 4 and 8), data fusion (Chapters 5 and 10), and data mining and analysis (Chapters 6 and 7). These methodologies are applied to address problems such as energy efficient driving control in a connected vehicle environment, traffic sensing data analysis and quality enhancement, TTR estimation, urban travel analysis, public transportation systems analysis, network

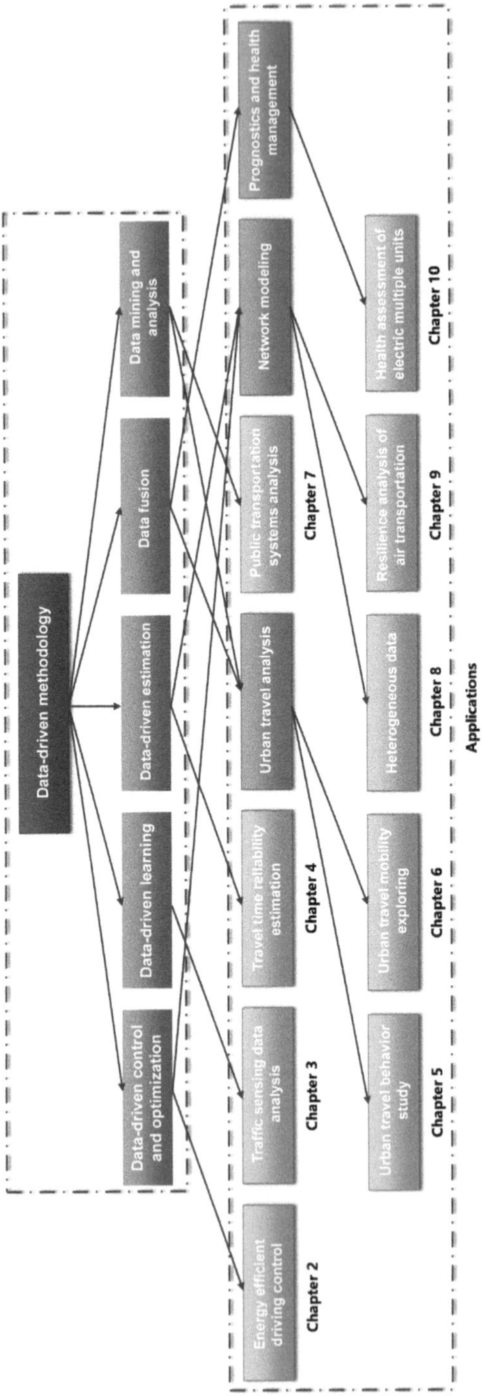

FIG. 1.2 A reader's guide to the structure and dependencies in this book.

modeling, and prognostics and health management. Specifically, management science-related topics, such as vehicle routing, network optimization, and information sharing, are also discussed in Chapters 5, 6, 8, and 10.

REFERENCES

[1] International Transport Forum, Data-Driven Transport Policy. Corporate Partnership Board Report, May. Organisation for Economic Co-operation and Development (OECD), 2016.

[2] The White House Office of the Press Secretary, FACT SHEET: Announcing Over $80 Million in New Federal Investment and a Doubling of Participating Communities in the White House Smart Cities Initiative, https://obamawhitehouse.archives.gov/the-press-office/2016/09/26/fact-sheet-announcing-over-80-million-new-federal-investment-and, 2016.

[3] O. Shijia, Baidu launches big data open platform to ease traffic, The 3rd World Internet Conference, China Daily, 2016. http://www.chinadaily.com.cn/business/3rdWuzhenWorldInternetConference/2016-11/18/content_27421197.htm.

[4] Transport Systems Catapult, Five-Year Delivery Plan to March 2018, (2013) https://ts.catapult.org.uk/wp-content/uploads/2016/04/Transport-Systems-Catapult-Five-Year-Delivery-Plan-to-March-2018.pdf.

[5] R. Cooper, Are We There Yet? Data-Driven Transportation on the Way, U.S. Chamber of Commerce Foundation, 2014. https://www.uschamberfoundation.org/blog/post/are-we-there-yet-data-driven-transportation-way/34417.

[6] J. Zhang, F. Wang, K. Wang, W. Lin, X. Xu, C. Chen, Data-driven intelligent transportation systems: a survey, IEEE Trans. Intell. Transp. Syst. 12 (4) (2011) 1624–1638.

[7] M. Nemschoff, Why the Transportation Industry Is Getting on Board With Big Data & Hadoop, MapR Technologies, 2014. https://mapr.com/blog/why-transportation-industry-getting-board-big-data-hadoop.

[8] D. Stone, R. Wang, Deciding With Data—How Data-Driven Innovation Is Fuelling Australia's Economic Growth, PricewaterhouseCoopers (PwC), Melbourne, 2014.

[9] Z. Cui, S. Zhang, K.C. Henrickson, Y. Wang, New progress of DRIVE net: an E-science transportation platform for data sharing, visualization, modeling, and analysis, Smart Cities Conference (ISC2), 2016 IEEE International, Trento, Italy, 2016, pp. 1–2.

[10] J. Cavanillas, E. Curry, W. Wahlster, New Horizons for a Data-Driven Economy, Springer, Berlin, 2016.

[11] S. Tabbitt, Big Data Analytics Keeps Dublin Moving, http://www.telegraph.co.uk/sponsored/sport/rugby-trytracker/10630406/ibm-big-dataanalytics-dublin.html, 2014.

[12] A. Khalid, T. Umer, M.K. Afzal, S. Anjum, A. Sohail, H.M. Asif, Autonomous data driven surveillance and rectification system using in-vehicle sensors for intelligent transportation systems (ITS), Comput. Netw. 139 (2018) 109–118.

[13] E.S. Rigas, S.D. Ramchurn, N. Bassiliades, Managing electric vehicles in the smart grid using artificial intelligence: a survey, IEEE Trans. Intell. Transp. Syst. 16 (4) (2015) 1619–1635.

[14] J. Jung, K. Sohn, Deep-learning architecture to forecast destinations of bus passengers from entry-only smart-card data, IET Intell. Transp. Syst. 11 (6) (2017) 334–339.

[15] C. Chen, H. Xiang, T. Qiu, C. Wang, Y. Zhou, V. Chang, A rear-end collision prediction scheme based on deep learning in the internet of vehicles, J. Parallel Distrib. Comput. 117 (2018) 192–204.

[16] M. Chowdhury, A. Apon, K. Dey, Data Analytics for Intelligent Transportation Systems, Elsevier, New York, 2017.

[17] D. Wei, Data-Driven Modeling and Transportation Data Analytics, (Ph.D. dissertation)Texas Tech University, 2014.

[18] F. Ghofrani, Q. He, R. Goverde, X. Liu, Recent applications of big data analytics in railway transportation systems: a survey, Transp. Res. Part C Emerg. Technol. 90 (2018) 226–246.

[19] H. Ghomi, M. Bagheri, L. Fu, L.F. Miranda-Moreno, Analysing injury severity factors at highway railway grade crossing accidents involving vulnerable road users: a comparative study, Traffic Inj. Prev. 17 (2016) 833–841.

[20] F. Shao, K. Li, X. Xu, Railway accidents analysis based on the improved algorithm of the maximal information coefficient, Intell. Data Anal. 20 (3) (2016) 597–613.

[21] J. Yin, W. Zhao, Fault diagnosis network design for vehicle on-board equipments of high-speed railway: a deep learning approach, Eng. Appl. Artif. Intell. 56 (October) (2016) 250–259.

[22] C. Hu, X. Liu, Modeling track geometry degradation using support vector machine technique, 2016 Joint Rail Conference. American Society of Mechanical Engineers, 2016 p. V001T01A011.

[23] B. Stratman, Y. Liu, S. Mahadevan, Structural health monitoring of railroad wheels using wheel impact load detectors, J. Fail. Anal. Prev. 7 (3) (2007) 218–225.

[24] L. Sun, Y. Lu, J.G. Jin, D.H. Lee, K.W. Axhausen, An integrated Bayesian approach for passenger flow assignment in metro networks, Transp. Res. Part C Emerg. Technol. 52 (2015) 116–131.

[25] S. Sharma, Y. Cui, Q. He, Z. Li, Data-driven optimization of railway track maintenance using Markov decision process, Proceedings of 96th Transportation Research Board Annual Meeting, Washington, DC, 2017.

[26] L. Kart, Advancing Analytics, (2013) 6. Online Presentation, April. Available from: http://meetings2.informs.org/analytics2013/Advancing%20Analytics_LKart_INFORMS%20Exec%20Forum_April%202013_final.pdf.

[27] X.L. Ma, Y.H. Wang, Development of a data-driven platform for transit performance measures using smart card and GPS data, J. Transp. Eng. 140 (12) (2014) 04014063.

[28] S. Tak, S. Kim, S. Oh, H. Yeo, Development of a data-driven framework for real-time travel time prediction, Comput. Aided Civ. Inf. Eng. 31 (10) (2016) 777–793.

[29] H. Perugu, H. Wei, Z. Yao, Integrated data-driven modeling to estimate PM2.5 pollution from heavy-duty truck transportation activity over metropolitan area, Transp. Res. Part D: Transp. Environ. 46 (2016) 114–127.

[30] S. Woo, S. Tak, H. Yeo, Data-driven prediction methodology of origin-destination demand in large network for real-time service, Transp. Res. Rec. 2567 (2016) 47–56.

[31] H. Khadilkar, Data-enabled stochastic modeling for evaluating schedule robustness of railway networks, Transp. Sci. 51 (4) (2017) 1161–1176.

[32] Z. Haider, A. Nikolaev, J.E. Kang, C. Kwon, Inventory rebalancing through pricing in public bike sharing systems, Eur. J. Oper. Res. 270 (1) (2018) 103–117.

Chapter 2

Data-Driven Energy Efficient Driving Control in Connected Vehicle Environment

Xuewei Qi*,†, Guoyuan Wu†, Kanok Boriboonsomsin† and Matthew J. Barth*,†

Department of Electrical and Computer Engineering, University of California, Riverside, CA, United States, †College of Engineering-Centre for Environmental Research and Technology (CE-CERT), University of California, Riverside, CA, United States

Chapter Outline

Data-Driven Solutions to Transportation Problems. https://doi.org/10.1016/B978-0-12-817026-7.00002-3

At the heart of Plug-in hybrid electric vehicles (PHEV) technologies, the energy management system (EMS) whose functionality is to control the power streams from both the internal combustion engine (ICE) and the battery pack based on vehicle and engine operating conditions have been studied extensively. In the past decade, a large variety of EMS implementations have been developed for HEVs and PHEVs, whose control strategies may be well categorized into two major classes:

(a) Rule-based strategies rely on a set of simple rules without a priori knowledge of driving conditions. Such strategies make control decisions based on instant conditions only and are easily implemented, but their solutions are often far from optimal due to the lack of consideration of variations in trip characteristics and prevailing traffic conditions.
(b) Optimization-based strategies are aimed at optimizing some predefined cost function according to the driving conditions and vehicle's dynamics. The selected cost function is usually related to the fuel consumption or tailpipe emissions.

Based on how the optimization is implemented, such strategies can be further divided into two groups: (1) off-line optimization which requires a full knowledge of the entire trip to achieve the global optimal solution; and (2) short-term prediction-based optimization, which takes into account the predicted driving conditions in the near future and achieves local optimal solutions segment by segment within an entire trip. However, major drawbacks of these strategies include heavy dependence on the knowledge of future driving conditions and high computational costs that are difficult to implement in real-time.

To address the aforementioned issues, we propose two data-driven on-line energy management strategies for PHEV energy efficient driving control in connected vehicle environment:

- Data-driven evolutionary algorithm-based self-adaptive EMS, which utilizes the rolling horizon technique to update the prediction of propulsion load as well as the power-split control. There are two major advantages over the existing strategies: (a) computationally competitive. There is no need to initiate a complete process for optimization while the algorithm keeps evolving and converging to obtain an optimal solution; (b) no a priori knowledge about the trip duration required.
- Data-driven reinforcement learning-based EMS, which is capable of simultaneously controlling and learning the optimal power-split operations in real-time from the historical driving data. There are three major features:

(1) this model can be implemented in real-time without any prediction efforts, since the control decisions are made only upon the current system state. The control decisions also considered for the entire trip information by learning the optimal or near-optimal control decisions from historical driving behavior. Therefore, this model achieves a good balance between real-time performance and energy saving optimality; (2) the proposed model is a data-driven model which does not need any PHEV model information once it is well trained, since all the decision variables can be observed and are not calculated using any vehicle powertrain models (these details are described in the following sections); and (3) compared to existing RL-based EMS implementations, the proposed strategy considers charging opportunities along the way (a key distinguishing feature of PHEVs as compared with HEVs).

The validation over real-world driving data has indicated that the proposed data-driving EMS strategies are very promising in terms of achieving a good balance between real-time performance and fuel savings when compared with some existing strategies, such as binary mode EMS and dynamic programming-based EMS. In addition, there is no requirement for the (predicted) information on the entire route.

2.1 INTRODUCTION

Air pollution and climate change impacts associated with the use of fossil fuels have motivated the electrification of transportation systems. In the realm of powertrain electrification, groundbreaking changes have been witnessed in the past decade in terms of research and development of hybrid electric vehicles (HEVs) and electric vehicles (EVs) [1]. As a combination of HEVs and EVs, PHEVs can be plugged into the electrical grid to charge their batteries, thus increasing the use of electricity and achieving even higher overall fuel efficiency, while retaining the ICE that can be called upon when needed [2].

In comparison to conventional HEVs, the EMS in PHEVs are significantly more complex due to their extended electric-only propulsion (or extended all-electric range capability) and battery chargeability via external electric power sources. Numerous efforts have been made in developing a variety of EMS for PHEVs [3, 4]. From the control perspective, existing EMS can be roughly classified as rule-based [5] and optimization-based [6]. This is discussed in more detail in Section 2.2.

In spite of all these efforts, most of the existing PHEV's EMS have one or more of the following limitations:

- Lack of adaptability to real-time information, such as traffic and road grade. This applies to rule-based EMS (either deterministic or using fuzzy logic) whose parameters or criteria have been pretuned to favor certain conditions (e.g., specific driving cycles and route elevation profiles) [3]. In addition, most EMS that are based on global optimization off-line assume that the future driving condition is known [2]. Thus far, only a few studies have focused on the development of on-line EMS for PHEVs [7].

- Dependence on accurate (or predicted) trip information that is usually unknown in advance. Many of the existing EMS require at a minimum the trip duration as known or predicted information prior to the trip [8]. Furthermore, it is reported that the performance of EMS is largely dependent on the time span of the trip [8]. Very few studies analyze the impacts of trip duration on the performance of EMS for PHEVs.
- Emphasis on a single trip level optimization without considering opportunistic charging between trips. The most critical feature that differentiates PHEVs from conventional HEVs is that PHEVs' batteries can be charged by plugging into an electrical outlet. Most of the existing EMS are designed to work on a trip-by-trip basis. However, taking into account inter-trip charging information can significantly improve the fuel economy of PHEVs [2].

2.2 BACKGROUND AND STATE OF THE ART

2.2.1 PHEV Modeling

Typically, there are three major types of PHEV powertrain architectures: (a) series, (b) parallel, and (c) power-split (series-parallel). This chapter focuses on the power-split architecture where the ICE and electric motors can power the vehicle, either alone or together, while the battery pack may be charged simultaneously through the ICE. Different approaches with various levels of complexity have been proposed for modeling PHEV powertrains [9]. However, a complex PHEV model with a large number of states may not be suitable for the optimization of PHEV energy control. A simplified but sufficiently detailed power-split powertrain model has been developed in MATLAB and used in this chapter. For more details, please refer to [2].

2.2.2 Operation Mode and SOC Profile

During the operation of a PHEV, the state-of-charge (SOC) may vary with time, depending on how the energy sources work together to provide the propulsion power at each instant. The SOC profile can serve as an indicator of the "PHEV" operating modes, i.e., charge sustaining (CS), pure electric vehicle (EV), and charge depleting (CD) modes [3], as shown in Fig. 2.1.

The CS mode occurs when the SOC is maintained at a certain level (usually the lower bound of SOC) by jointly using power from both the battery pack and the ICE. The pure EV mode is when the vehicle is powered by electricity only. The CD mode represents the state when the vehicle is operated using power primarily from the battery pack with supplemental power from the ICE as necessary. In the CD mode, the ICE is turned on if the electric motor is not able to

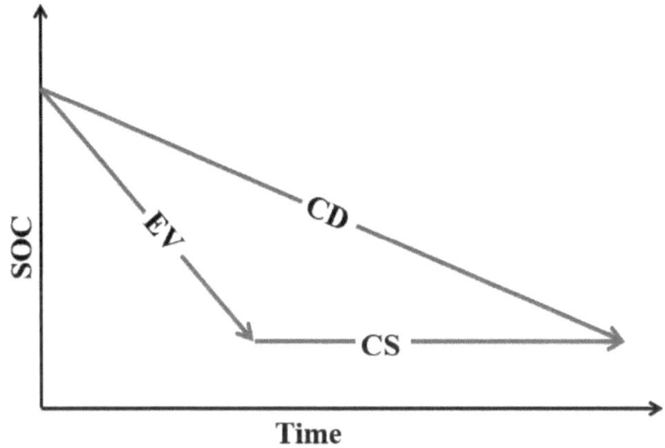

FIG. 2.1 Basic operation modes for PHEV.

provide enough propulsion power or the battery pack is being charged (even when the SOC is much higher than the lower bound) in order to achieve better fuel economy.

2.2.3 EMS for PHEVs

The goal of the EMS in a PHEV is to satisfy the propulsion power requirements while maintaining the vehicle's performance in an optimal way. A variety of strategies have been proposed and evaluated in many previous studies [4]. A detailed literature review on EMS for PHEVs is provided in this section. Broadly speaking, the existing EMS for PHEVs can be divided into two major categories:

(1) Rule-based EMS are fundamental control schemes operating on a set of predefined rules without prior knowledge of the trip. The control decisions are made according to the current vehicle states and power demand only. Such strategies are easily implemented, but the resultant operations may be far from being optimal due to not considering future traffic conditions.
(2) Optimization-based EMS aim at optimizing a predefined cost function according to the driving conditions and behaviors. The cost function may include a variety of vehicle performance metrics, such as fuel consumption and tailpipe emissions.

For rule-based EMS, deterministic and fuzzy control strategies (e.g., binary control) have been well investigated. For optimization-based EMS, the strategies can be further divided into three subgroups based on how the optimizations are implemented:

(1) Off-line strategy which requires a full knowledge of the entire trip beforehand to achieve the global optimal solution;

(2) Prediction-based strategy or so-called real-time control strategy which takes into account predicted future driving conditions (in a rolling horizon manner) and achieves local optimal solutions segment-by-segment. This group of strategies is quite promising due to the rapid advancement and massive deployment of sensing and communication technologies (e.g., GPS) in transportation systems that facilitate the traffic state prediction; and.
(3) Learning-based strategy which is recently emerging owing to the research progress in machine learning techniques. In such a data-driven strategy, a dynamic model is no longer required. Based on massive historical and real-time information, trip characteristics can be learned and the corresponding optimal control decisions can be made through advanced data mining schemes. This strategy fits very well for commute trips.

Fig. 2.2 presents a classification tree of EMS for PHEVs and the typical strategies in each category, based on most existing studies.

In addition to the classification above, Table 2.1 highlights several important features which help differentiate the aforementioned strategies. Example references are also included in Table 2.1.

2.2.4 PHEVs' SOC Control

For a power-split PHEV, the optimal energy control is, in principle, equivalent to the optimal SOC control. Most of the existing EMS for PHEVs implicitly integrate SOC into the dynamic model and regard it as a key control variable [25], while only a few studies have explicitly described their SOC control strategies. A SOC reference control strategy is proposed in [20] where a supervisory SOC

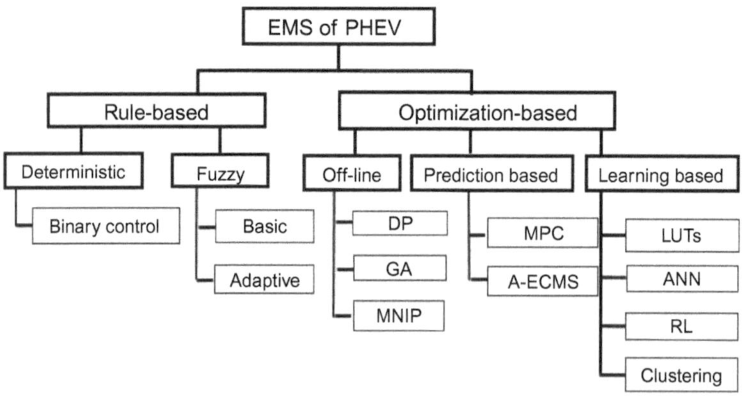

FIG. 2.2 Basic classification of EMS for PHEV. *Note*: PMP, Pontraysgin's minimum principle; MNIP, mixed nonlinear integer programming; DP, dynamic programming; QP, quadratic programming; RL, reinforcement learning; ANN, artificial neural network; LUTs, look-up-tables; MPC, model predictive control; AECMS, adaptive equivalent consumption minimization strategy.

TABLE 2.1 Classification of Current Literature

	Rule-Based	Off-Line Optimization	Prediction-Based	Learning-Based
Optimality	Local	Global	Local	Local
Real-time	Yes	No	Yes	Yes
SOC control	No	Yes	Yes	No
Need trip duration	No	Yes	Yes	Yes
Example references	[7,10–12]	[2, 6, 13–17]	[8, 18–23]	[9, 18, 19, 24–26]

planning method is designed to precalculate an optimal SOC reference curve. The proposed EMS then tries to follow this curve during the trip to achieve the best fuel economy. Another SOC control strategy is proposed in [8], where a probabilistic distribution of trip duration is considered. More recently, machine learning-based SOC control strategies (e.g., [9]) have emerged, where the optimal SOC curves are precalculated using historical data and stored in the form of look-up tables for real-time implementation. A common drawback for all these strategies is that accurate trip duration information is required in an either deterministic or probabilistic way. In reality, however, such information is hard to know ahead of time or may vary significantly due to the uncertainties in traffic conditions. To ensure the practicality of our proposed EMS for PHEVs, we employ a self-adaptive SOC control strategy in this chapter that does not require any information about the trip duration (or length).

2.3 PROBLEM FORMULATION

2.3.1 Data-Driven On-Line EMS Framework for PHEVs

In this chapter, we propose an on-line EMS framework for PHEVs, using the receding horizon control structure (see Fig. 2.3). The proposed EMS framework consists of information acquisition (from external sources), prediction, optimization, and power-split control. With the receding horizon control, the entire trip is divided into segments or time horizons. As shown in Fig. 2.4, the prediction horizon (N sampling time steps) needs to be longer than the control horizon (M sampling time steps). Both horizons keep moving forward (in a rolling horizon style) while the system is operating. More specifically, the prediction model is used to predict the power demand at each sampling step (i.e., each second) in the prediction horizon. Then, the optimal ICE power supply for each second during the prediction horizon is calculated with this predicted information.

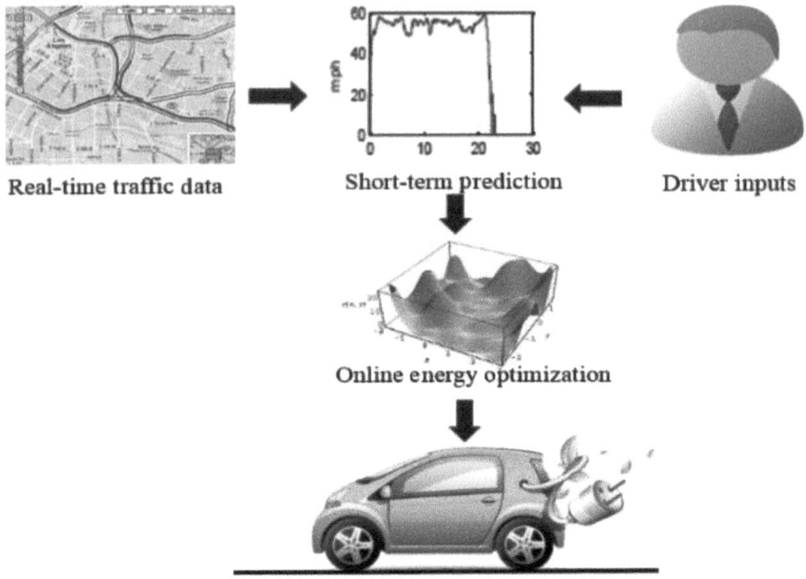

FIG. 2.3 Flow chart of the proposed on-line EMS.

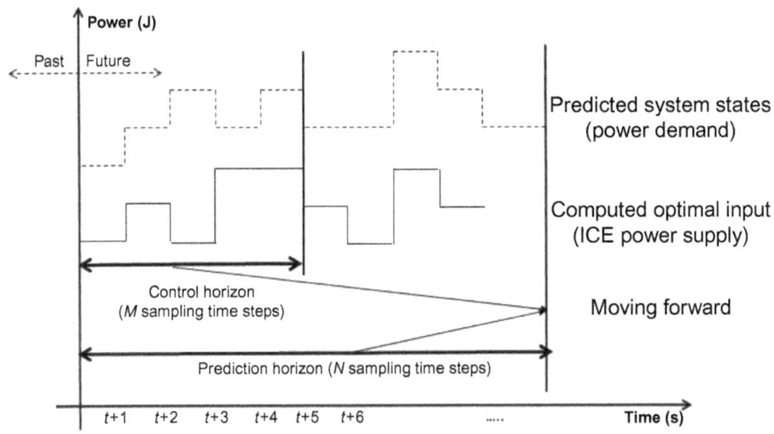

FIG. 2.4 Time horizons of prediction and control.

In each control horizon, the precalculated optimal control decisions are inputted into the powertrain control system (e.g., electronic control unit, or ECU) at the required sampling frequency. In this chapter, we focus on the on-line energy optimization, assuming that the short-term prediction model is available (which is one of our future research topics).

2.3.2 Optimal Power-Split Control Formulation

Mathematically, the optimal (in terms of fuel economy) energy management for PHEVs can be formulated as a nonlinear constrained optimization problem. The objective is to minimize the total fuel consumption by ICE along the entire trip.

$$\begin{cases} \min\left\{\int_0^T h(\omega_e, q_e, t)dt\right\} \\ \text{subject to}: \\ \dot{\text{SOC}} = f(\text{SOC}, \omega_{\text{MG1}}, q_{\text{MG1}}, \omega_{\text{MG2}}, q_{\text{MG2}}) \\ (\omega_e, q_e) = g(\omega_{\text{MG1}}, q_{\text{MG1}}, \omega_{\text{MG2}}, q_{\text{MG2}}) \\ \text{SOC}_{\min} \le \text{SOC} \le \text{SOC}_{\max} \\ \omega_{\min} \le \omega_e \le \omega_{\max} \\ q_{\min} \le q_e \le q_{\max} \end{cases} \tag{2.1}$$

where T is the trip duration, ω_e, q_e are the engine's angular velocity and engine's torque, respectively, $h(\omega_e, Tq_e)$ is ICE fuel consumption model, ω_{MG1}, q_{MG1} are the first motor/generator's angular velocity and torque, respectively, ω_{MG2}, q_{MG2} are the second motor/generator's angular velocity and torque, respectively, and $f(\text{SOC}, \omega_{\text{MG1}}, q_{\text{MG1}}, \omega_{\text{MG2}}, q_{\text{MG2}})$ is the battery power consumption model. For more details about the model derivations and equations, please refer to [2].

Such a formulation is quite suitable for traditional mathematical optimization methods [13] with high computational complexity. In order to facilitate on-line optimization, we herein discretize the engine power and reformulate the optimization problem represented by Eq. (2.1) as follows:

$$\min \sum_{k=1}^{T} \sum_{i=1}^{N} x(k,i) P_i^{\text{eng}} / \eta_i^{\text{eng}} \tag{2.2}$$

subject to

$$\sum_{k=1}^{j} f\left(P_k - \sum_{i=1}^{N} x(k,i) P_i^{\text{eng}}\right) \le C \quad \forall j = 1, \ldots, T \tag{2.3}$$

$$\sum_{i=1}^{N} x(k,i) = 1 \quad \forall k \tag{2.4}$$

$$x(k,i) = \{0, 1\} \quad \forall k, i \tag{2.5}$$

where N is the number of discretized power level for the engine, k is the time step index, i is the engine power level index, C is the gap of the battery pack's SOC between the initial and the minimum, P_i^{eng} is the ith discretized level for the engine power and η_i^{eng} is the associated engine efficiency, and P_k is the driving power demand at time step k.

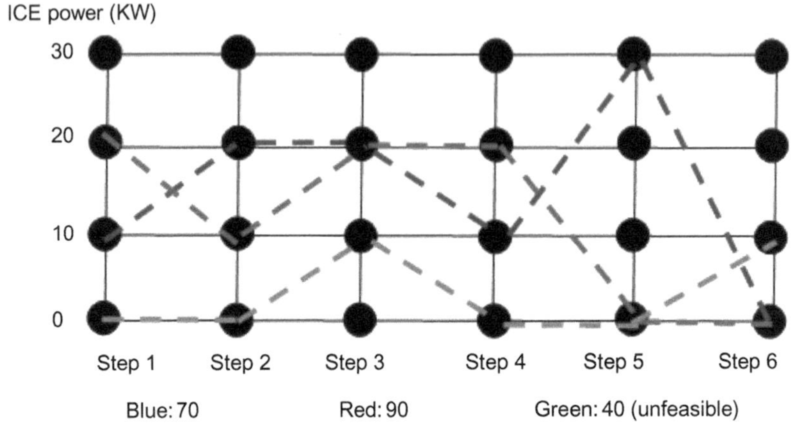

FIG. 2.5 Example solutions of power-split control.

Furthermore, if the change in SOC (ΔSOC) for each possible engine power level at each time step is pre-calculated given the (predicted) power demand, then constraint (2.3) can be replaced by

$$\mathrm{SOC}^{\mathrm{ini}} - \mathrm{SOC}^{\mathrm{max}} \leq \sum\nolimits_{k=1}^{j} x(k,i)\Delta\mathrm{SOC}(k,i) \leq \mathrm{SOC}^{\mathrm{ini}} - \mathrm{SOC}^{\mathrm{min}} \quad \forall j = 1,\ldots,T \tag{2.6}$$

where $\mathrm{SOC}^{\mathrm{ini}}$ is the initial SOC, and $\mathrm{SOC}^{\mathrm{min}}$ and $\mathrm{SOC}^{\mathrm{max}}$ are the minimum and maximum SOC, respectively. Therefore, the problem is turned into a combinatory optimization problem whose objective is to select the optimal ICE power level for each time step given the predicted information in order to achieve the highest fuel efficiency for the entire trip. Fig. 2.5 gives three example ICE power output solutions. The solution represented by the blue line (starting from 20 KW) has a lower total ICE power consumption (i.e., 40 units) than the red line (starting from 10 KW) (i.e., 90 units), while the green line (starting from 0 KW) represents an infeasible solution due to the SOC constraint.

2.4 DATA-DRIVEN EVOLUTIONARY ALGORITHM (EA) BASED SELF-ADAPTIVE ON-LINE OPTIMIZATION

The motivations for applying EA are:

(1) compared to the traditional derivative or gradient-based optimization methods, EAs are easier to implement and require less complex mathematical models;
(2) EAs are very good at solving nonconvex optimization problems where there are multiple local optima; and
(3) it is very flexible to address multiobjective optimization problems using EAs.

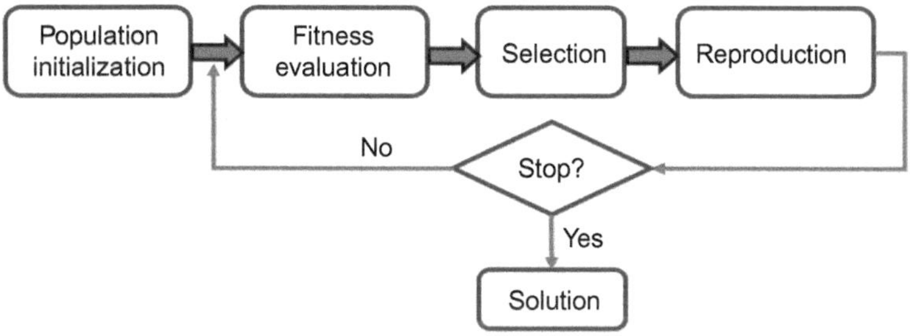

FIG. 2.6 Estimation and sampling process of EA.

Theoretically, in the proposed framework, any EAs can be used to solve the optimization problem for each prediction horizon described in Fig. 2.4. A typical EA is a population-based and iterative algorithm that starts searching for the optimal solution with a random initial population. Then, the initial population undergoes an iterative process that includes multiple operations, such as fitness evaluation, selection, and reproduction, until certain stopping criteria are satisfied. The flow chart of an EA is provided in Fig. 2.6.

Among many EAs, the estimation distribution algorithm (EDA) is very powerful in solving high-dimensional optimization problems and has been applied successfully to many different engineering domains [27]. In this chapter, we choose EDA as the major EA kernel in the proposed framework due to the high-dimensionality nature of the PHEV energy management problem. This selection is justified by experimental results in the following sections.

In the problem representation of EDA, each individual (encoded as a row vector) of the population defined in the algorithm is a candidate solution. For the PHEV energy management problem, the size of the individual (vector) is the number of time steps within the trip segment. The value of the ith element of the vector is the ICE power level chosen for that time step. In the example individual in Table 2.2, the ICE power level is 3 (or 3 kW) for the first time step, 0 kW (i.e., only battery pack supplies power) for the second time step, 1 for the third time step, and so forth.

It is very flexible to define a fitness function for EAs. Since the objective is to minimize fuel consumption, the fitness function herein can be defined as the summation of total ICE fuel consumption for the trip segment defined by Eq. (2.5) and a penalty term

$$f(s) = C_{\text{fuel}} + P \tag{2.7}$$

where s is a candidate solution, C_{fuel} is fuel consumption, and P is the imposed penalty that is the largest possible amount of energy that can be consumed in this trip segment. The penalty is introduced to guarantee the feasibility of the solution, satisfying constraint (2.3), which means that the SOC should always

TABLE 2.2 Representation of One Example Individual

Time	1 s	2 s	3 s	4 s	....	$n-3$	$n-2$	$n-1$	n
Individual	3	0	1	4	....	1	2	0	5

fall within the required range at each time step. Then, all the individuals in the population are evaluated by the fitness function and ranked by their fitness values in an ascending order since this is a minimization problem. A good evaluation and ranking process is crucial in guiding the evolution towards good solutions until the global optima (or near optima) is located.

Furthermore, EDA assumes that the value of each element in a good individual of the population follows a univariate Gaussian distribution. This assumption has been proven to be effective in many engineering applications [28], although there could be other options [29]. For each generation, the top individuals (candidate solutions) with least fuel consumption values are selected as the parents for producing the next generation by an estimation and sampling process [30].

The flow chart of the proposed EDA-based on-line EMS is presented in Fig. 2.7. t_0 is the current time, N is the length of the prediction time horizon,

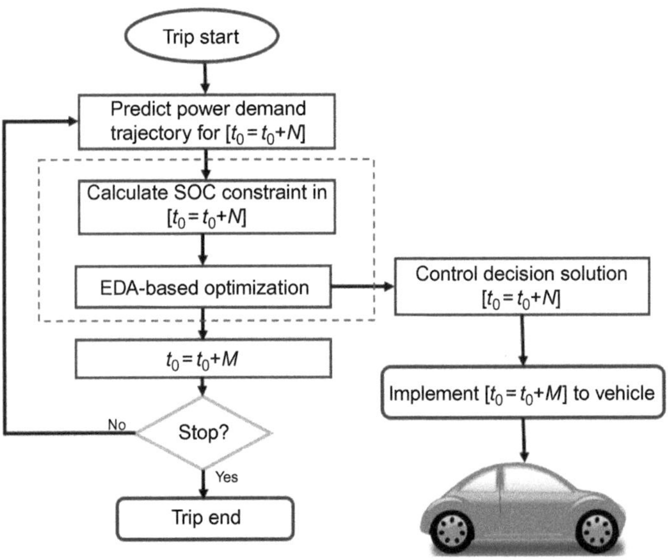

FIG. 2.7 EDA-based on-line energy management system.

and M is length of the control time horizon. The block highlighted by the dashed box is the core component of the system, and more details about this block is given in Section 2.4.

2.4.1 Optimality and Complexity

Evolutionary algorithms (EA) are stochastic search algorithms that do not guarantee to find the global optima. Hence, in the proposed on-line EMS, the optimal power control for each trip segment is not guaranteed to be found. Moreover, EAs are also population-based iterative algorithms that are usually criticized due to their heavy computational loads [31], especially for real-time applications. Theoretically, time complexity of EAs is worse than $\theta(m^2 * \log(m))$ where m is the size of the problem [32]. However, we apply the receding horizon control technique in this chapter, where the entire trip is divided into small segments. Therefore, the computational load can be significantly reduced since the EA-based optimization is applied only for each small segment rather than the entire trip. In this sense, the proposed framework can be implemented in "real-time," as long as the optimization for the next prediction horizon can be completed in the current control horizon (see Fig. 2.4). As previously discussed, the rule-based EMS can run in real-time but the results may be far from optimal while most of the optimization-based EMS have to operate off-line. Therefore, the proposed on-line EMS would be a well-balanced solution between the real-time performance and optimality.

2.4.2 SOC Control Strategies

An appropriate SOC control strategy is critical in achieving the optimal fuel economy for PHEVs [33]. In the previously presented problem formulation, the major constraint for SOC is defined by Eq. (2.6), which means that at any time step, the SOC should be within the predefined range (e.g., between 0.2 and 0.8) to avoid damage to the battery pack. However, this constraint only may not be enough to accelerate the search for the optimal solution. Hence, additional constraint(s) on battery use (e.g., reference bound of SOC) should be introduced to improve the on-line EMS. To investigate the effectiveness of different SOC control strategies within the proposed framework, two types of SOC control strategies—reference control and self-adaptive control—are designed and evaluated in this chapter.

2.4.2.1 SOC Reference Control (Known Trip Duration)

When the trip duration is known, a SOC curve can be pre-calculated and used as a reference to control the use of battery power along the trip to achieve optimal fuel consumption. We propose three heuristic SOC references (i.e., lower

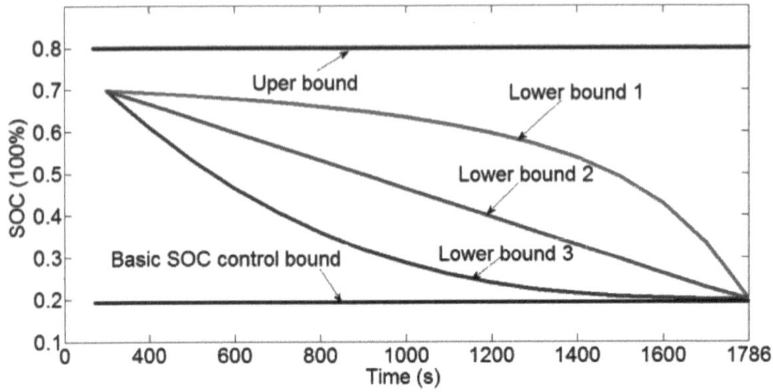

FIG. 2.8 SOC reference control bound examples.

bounds) in this chapter (see Fig. 2.8 for example): (1) concave downward; (2) straight line; and (3) concave upward. These SOC minimum bounds are generated based on the given trip duration information by the following equations, respectively:

- Concave downward control (lower bound 1):

$$SOC_i^{min} = \frac{\left(SOC_{init} - SOC^{min}\right)}{T-(i^*M)} {}^*N + SOC^{init} \tag{2.8}$$

- Straight line control (lower bound 2):

$$SOC_i^{min} = \frac{-\left(SOC_i^{min} - SOC^{min}\right)}{T} \cdot ((i-1)\cdot M + N) + SOC^{init} \tag{2.9}$$

- Concave upward control (lower bound 3):

$$SOC_i^{min} = \frac{-\left(SOC_{i-1}^{end} - SOC^{min}\right)}{T-(i^*M)} {}^*N + SOC_{i-1}^{end} \tag{2.10}$$

where i is the segment index; $SOC_i{}^{min}$ is the minimum SOC at the end of ith segment; and $SOC_{i-1}{}^{end}$ is the SOC at the end of last control horizon. It is self-evident that the concave downward bound (i.e., lower bound 1) is much more restrictive than a concave upward bound (i.e., lower bound 3) in terms of battery energy use at the beginning of the trip.

A major drawback for these reference control strategies is that they assume that the trip duration (i.e., T) is given, or at least can be well estimated beforehand. As mentioned earlier, this assumption may not hold true for many real-world applications. Therefore, a new SOC control strategy without relying on the knowledge of trip duration would be more attractive.

TABLE 2.3 Example Fitness Evaluation by Different Fitness Functions

Indiv. Index	Fuel Con.	SOC Decrease	R_{fuel}	R_{soc}	Rank by Eq. (2.7)	Rank by Eq. (2.11)
1	0.001	0.005(P)	5	35	98	140
2	0.010	0.002	25	14	33	39
3	0.007	0.003	19	23	24	42
4	0.002	0.004(P)	7	32	99	139
....	......	....	....		....	....

2.4.2.2 SOC Self-Adaptive Control (Unknown Trip Duration)

In this chapter, we also propose a novel self-adaptive SOC control strategy for real-time optimal charge-depleting control, where trip duration information is not required. Unlike those SOC reference control strategies that control the use of battery by explicit reference curves, the self-adaptive control strategy controls the battery power utilization implicitly by adopting a new fitness function in place of the one in Eq. (2.7):

$$f(s) = R_{\text{fuel}} + R_{\text{soc}} + P' \tag{2.11}$$

where R_{fuel} and R_{soc} are the ranks (in an ascending order) of ICE fuel consumption and SOC decrease, respectively, of an individual candidate solution s in the current population; and $P'$is the added penalty when the individual s violates the constraints given in Eq. (2.6). The penalty value is selected to be greater than the population size in order to guarantee that an infeasible solution always has a lower rank (i.e., larger fitness value) than a feasible solution in the ascending order by fitness value. Compared to the fitness function adopted for SOC reference control (see Eq. (2.7)), this new fitness function tries to achieve a good balance between two conflicting objectives: least fuel consumption and least SOC decrease. For a better understanding of the differences between these two fitness functions, Table 2.3 provides an example of fitness evaluation of the same population. In this case, the population size is 100. As we can see in the table, Individual 2, who has a better balance between fuel consumption, and SOC decrease, is more favorable than Individual 3 in the ranking by Eq. (2.11) than that by Eq. (2.7).

2.4.3 EDA-Based On-Line EMS Algorithm With SOC Control

Details of the proposed EDA-based on-line EMS algorithm with SOC control are summarized in Algorithm 1. This algorithm is implemented on each prediction horizon (N time steps) within the framework presented in Fig. 2.8 (see the box with dashed line).

Algorithm 1: EDA-based on-line EMS with SOC control

1: Initialize a random output solution I_{best}(N time steps)
2: $P_{current}$ <= Generate initial population randomly
3: While iteration_number ≤ Max_iterations, do
4: For each individual s in $P_{current}$
5: Calculate fuel consume C_{fuel} using Eq. (2.1).
6: Calculate SOC decrease using Eq. (2.5)
7: Obtain the rank index of s: R_{fuel}
8: Obtain the rank index of s: R_{soc}
9: If SOC reference control is adopted
10. Calculate the lower bound using Eqs. (2.8)–(2.10)
11: If individual s violates Eq. (2.6)
12: $P = P_0$;//largest fuel consumption in N steps
13: Else
14: $P = 0$;
15: End If
16: Calculate the fitness value for s using Eq. (2.7)
17: Else If SOC self-adaptive control is adopted
18: If individual s violates Eq. (2.6)
19: $P' = S$
20: Else
21: $P' = 0$;
22: End If
23: Calculate the fitness value for s using Eq. (2.11)
24: End If
25: End For
26: Rank $P_{current}$ in ascending order based on fitness
27: P_{top} <= Select top α% individuals from $P_{current}$
28: E < = Estimate a new distribution from P_{top}
29: P_{new} <= Sample N individuals from built model E
30: Evaluate each individual in P_{new} using line 5–14
31: Mix $P_{current}$ and P_{new} to form 2N individuals
32: Rank 2N individuals in ascending order by fitness
33: $P_{current}$<= Select top N individuals
34: Update I_{best} if a better one is identified.
35: Iteration_number ++
36: End While
37: Output I_{best}

In the following section, we compare the performance of the proposed self-adaptive SOC control with other SOC control strategies. For convenience, we list the abbreviations of all the involved strategies in Table 2.4.

TABLE 2.4 Abbreviations of Different SOC Control Strategies Compared in This Chapter

SOC Control Strategies	Abbreviations
Binary control	B-I
Basic SOC control	B-A
Concave downward	C-D
Straight line	S-L
Concave upward	C-U
Self-adaptive SOC control	S-A

2.4.4 Synthesized Trip Information

To validate the proposed EMS for PHEVs, we use real-world data collected on January 17, 2012, along I-210 between I-605 and Day Creek Blvd in San Bernardino, California, as a case study (see Fig. 2.9). Please refer to [2] for more detailed description of data collection and specifications of the power-split PHEV model if interested.

Based on the collected traffic data along with road grade information, second-by-second vehicle velocity trajectory and power demand have been synthesized as described in [2]. As pointed out earlier, it is impractical to have a priori knowledge of the exact vehicle velocity trajectory. In this chapter, we focus on the development of the optimal power-split control, assuming perfect

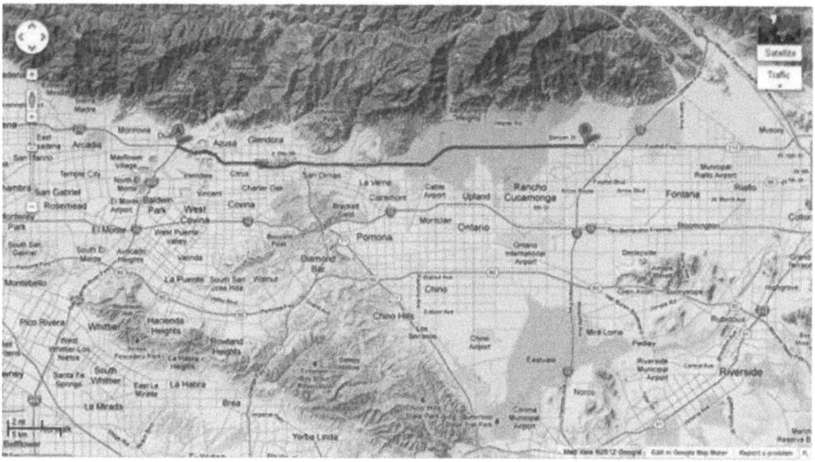

FIG. 2.9 Example trip along I-210 in southern California used for evaluation.

prediction of vehicle velocity trajectory. Research on improving the prediction of vehicle velocity trajectory in real-time is part of our future work.

2.4.5 Off-Line Optimization for Validation

To justify the selection of EDA as the kernel of the proposed framework, we first test EDA on the full-trip off-line optimization. The results are compared with those obtained from two other popular EA: genetic algorithm (GA) and particle swarm optimization (PSO). The fitness (i.e., total ICE energy consumption) of EDA-based off-line optimization obtains better fuel economy (0.346 gal) than the other two (0.364 gal for GA and 0.377 for PSO, respectively), at the same computational expense (i.e., same population size and same number of iterations) [30]. In addition, the result from EDA is much closer to the global optimum (0.345 gal in this case), with the difference being $<1\%$.

2.4.6 Real-Time Performance Analysis and Parameter Tuning

As aforementioned, a necessary condition for on-line implementation of the proposed EMS is that the optimization for the next prediction horizon has to be finished within the current control horizon (see Fig. 2.4). In our study, for example, the optimization for a prediction horizon of 50 s can be completed within 1.1 s (with Intel Core i7, 3.4 GHz, RAM 4G, and 64bit-Matlab 2012). In addition, one of our previous works [30] has shown that the lengths of prediction horizon and control horizon may significantly affect the algorithm performance. The best combination of these two parameters is found to be $N = 250$ and $M = 10$ in this case.

Unlike the conventional MPC, whose optimization has to be implemented along each prediction horizon, our proposed EA-based online EMS (see Fig. 2.7) can take advantage of the optimal results from previous prediction horizons, which avoids a new optimization starting from scratch and therefore saves a lot of computational overheads. As can be seen in Fig. 2.10, part of the optimal decisions from previous prediction optimization horizon are adopted as the seed for initial population of current prediction horizon optimization. For example, when the control horizon is 3 s and prediction/optimization horizon

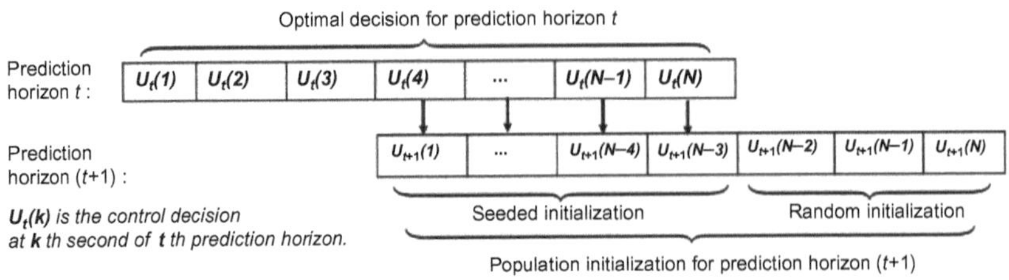

FIG. 2.10 Population initialization from the second prediction horizon (i.e., $t \geq 2$).

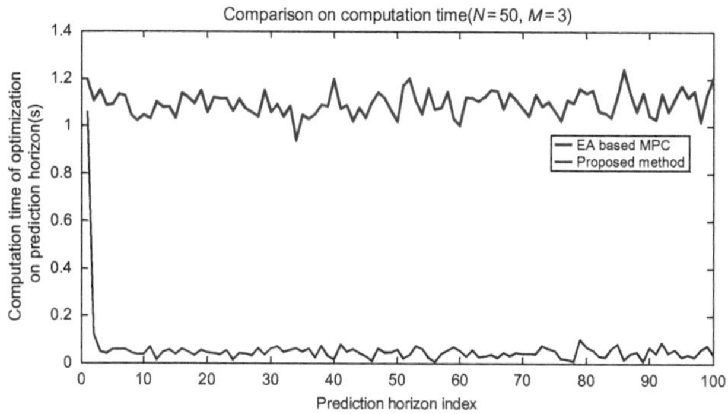

FIG. 2.11 Comparison of computation time.

is N, only 3 control decisions need to be randomly initialized and optimized in the second prediction/optimization horizon. This allows the optimization or search to be much more efficient compared to the same process over entire prediction horizon. To validate further this computational performance, we designed an EA-based MPC (EAMPC) that activates a complete new optimization for each prediction/optimization horizon, and compared it with our proposed model. The computation time track in Fig. 2.11 shows that for a 50-s prediction horizon, the conventional MPC takes around 1.1 s for each optimization horizon, but our proposed model can take <0.1 s to finish the optimization from the second prediction horizon.

2.4.7 On-Line Optimization Performance Comparison

To evaluate fully the performance of the proposed on-line EMS strategies, we compare them to the conventional binary control (implementable in real-time) strategy as well as the off-line global optimal control strategy (with the use of dynamic programming (DP) [11]). The comparisons are carried out on both the single trip scenario and multiple trips scenario.

When tested on a single (westbound) trip, the fuel consumption and SOC profiles by different strategies are illustrated in Fig. 2.12. It is shown that the proposed S-A algorithm achieves the lowest fuel consumption (0.3515 gal) which is only 1.56% worse than that of global optima obtained by the off-line optimization (0.3460 gal). These results can be explained by the shape of the resultant SOC profiles. For instance, SOC decreases very quickly in the B-I strategy, and reaches the lower bound (i.e., 0.2) at around 1200 s because the use of battery power is always prioritized whenever available. Therefore, ICE has to supply most of the demanded power after 1200 s. This is very similar to the cases of the B-A and C-U strategies where the battery power is also consumed aggressively at the beginning of the trip with very loose constraints. On the other hand, the S-L and C-D strategies perform better since their battery

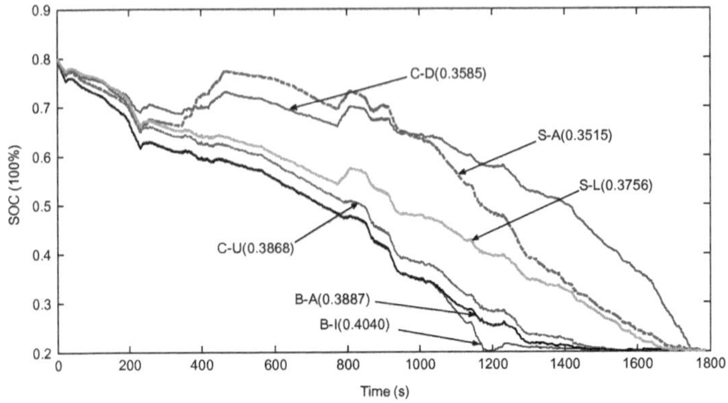

FIG. 2.12 SOC trajectories resulted from different control strategies.

power is used more cautiously along the trip. These findings are consistent with the conclusions of many other studies [8, 33] in that a smoother distribution of battery power usage along the trip would result in higher fuel efficiency.

In order to understand the statistical significance of the different EMS strategies, we test them on 30 randomly selected trip profile data extracted from the same road segment on 12 different days. The results are also compared to the binary control and D-P strategies. For the purpose of comparison, we set the fuel consumption obtained by the binary control strategy as the baseline and calculate the percentage of fuel savings achieved by the other EMS strategies. As we can see in Fig. 2.13, the D-P strategy achieves the best fuel savings with an average of 19.4% and the least variance simply because it is an off-line optimization strategy. The proposed S-A strategy achieves an average of 10.7% fuel savings, which is higher than all other on-line strategies and consistent with the result of

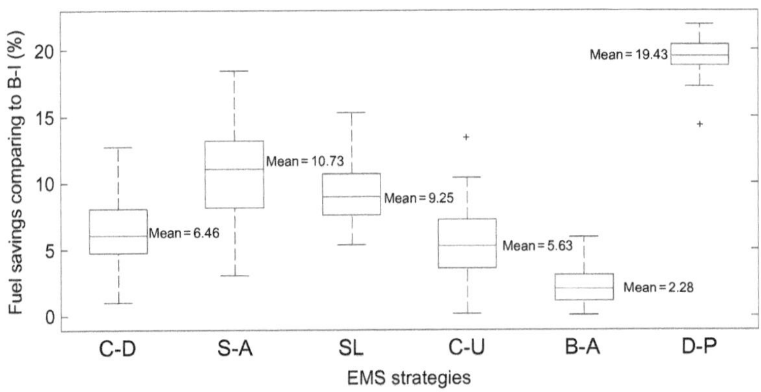

FIG. 2.13 Box-plot of fuel savings on 30 trips.

TABLE 2.5 Comparisons With Existing Models

EMS Model	Year	STP[a]	Trip Distance	FEI[b]	Consider Charging
This work	2016	Yes	Unknown	10.7%	Yes
EAMPC	2016	Yes	Unknown	7.9%	Yes
MPC [21]	2014	Yes	Known	8.5%	No
MPC [20]	2015	Yes	Known	6.7%	No
A-ECMS [21]	2014	Yes	Known	10.2%	No
A-ECMS [18]	2015	Yes	Known	7.6%	No
DP [22]	2015	Yes	Known	5.8%	No
SDP[c] [23]	2011	Yes	Known	7.7%	No

[a]*Short-term prediction.*
[b]*Fuel economy improvement comparing to binary control.*
[c]*Stochastic dynamic programming.*

the single trip test. An interesting observation is that the S-L strategy has better average fuel savings (i.e., 9.3%) than the C-D and C-U strategies, which is not consistent with the test result of the single trip test. A possible reason is that the C-D strategy performs better on some trips in which the power demand is higher in later stages of the trip, but the C-U strategy performs better on the trips in which the power demand is higher in earlier stages. On the other hand, the S-L strategy balances the SOC control between these two types of trip pattern, and therefore has better average performance.

For further validation, the proposed S-A strategy with the best performance is compared with other existing PHEV EMS strategies that employ short-term prediction. Although these strategies were proposed to handle powertrain models with different fidelity as well as different data set for validation, they all used the binary control strategy as a benchmark (the same as used in this chapter). This provides us with a chance to compare all models in a relatively fair manner. The comparison results are listed in Table 2.5, which proves that our model achieves the largest improvement of fuel efficiency (with regard to the binary control strategy) but requires less trip information.

2.4.8 Analysis of Trip Duration

In this section, we analyze and compare the effectiveness of the proposed on-line EMS for longer trips. These longer trips are constructed by concatenating multiple trip profiles, and the results are shown in Fig. 2.14. As can be

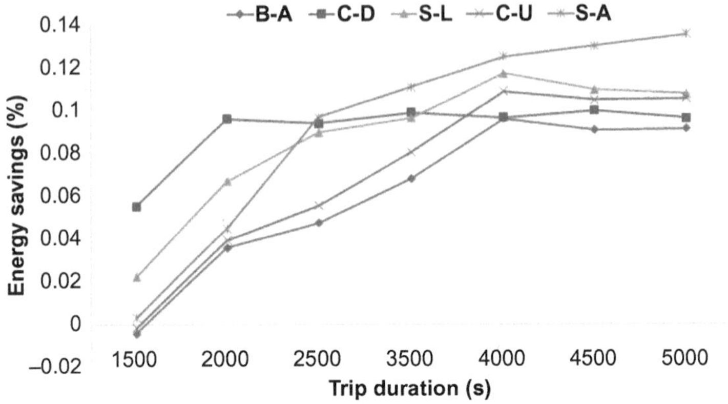

FIG. 2.14 Fuel savings for trips with different duration, compared to B-I.

observed, the B-I strategy has the best fuel economy when the trip duration is shorter than 1500 s. For these short trips, the PHEV can mostly rely on battery energy. However, as the trip duration becomes longer, especially when longer than 2500 s, the S-A strategy outperforms all the others.

To explain this finding further, the resultant fuel consumption and the corresponding SOC profiles for the longest trip (5000 s) are provided in Fig. 2.15. According to the figure, the S-A strategy has the lowest fuel consumption and its SOC profile is a combination of the CD mode (defined in Fig. 2.1) before 2000 s and the CS mode after 2000 s. This contradicts with most of the existing studies, which report that an optimal fuel economy for the trip can be achieved by operating solely in the CD mode [8]. Here, we present evidence that it is not always the case, and that the CD + CS operation can result in optimal fuel

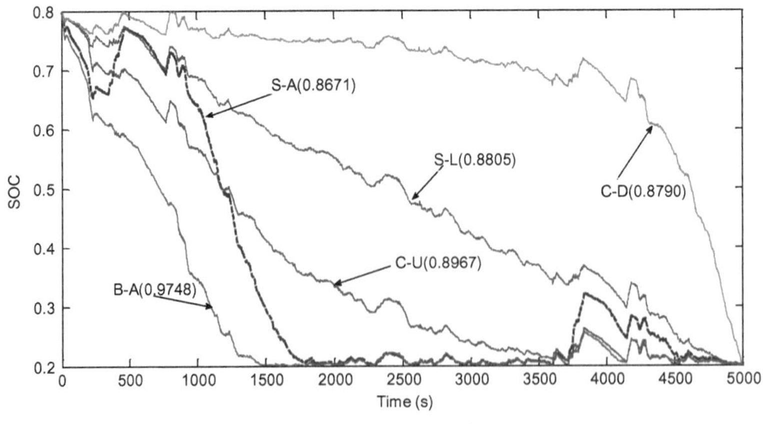

FIG. 2.15 Resultant SOC curve when trip duration is 5000 s.

efficiency for long trips. Furthermore, this finding also implies the potential for the proposed S-A strategy to adapt to different trip durations.

2.4.9 Performance With Charging Opportunity

Considering the plug-in capability of PHEVs, we evaluate the performance of the proposed strategies at the tour level. More specifically, we consider the commute trips of the case study as a tour and assume that there is a charging opportunity (to a full charge) between the end of the westbound trip and the beginning of the eastbound trip. We then compare the different SOC control strategies under the following two scenarios:

(1) Scenario I: The proposed EMS with a priori knowledge of the charging opportunity;
(2) Scenario II: The proposed EMS without a priori knowledge of the charging opportunity. In this case, a conservative strategy is applied by assuming that there is no charging station available in between the trips.

The results are illustrated in Fig. 2.16. They show that the knowledge of the charging opportunity information has great influence on the resultant SOC profiles for the deterministic SOC reference control strategies but no influence on the SOC self-adaptive control strategy. Table 2.6 presents the increased fuel consumption due to the lack of knowledge of the charging opportunity prior to the tour. As shown in the table, the C-D, S-L, and C-U strategies all have 13% or more increase in fuel consumption if the charging opportunity information is unknown, while the B-I and S-A strategies are not affected because the trip duration is not considered in their decision-making process. These findings further emphasize the advantage of the proposed SOC self-adaptive control strategy in terms of robustness to the level of knowledge about charging availability.

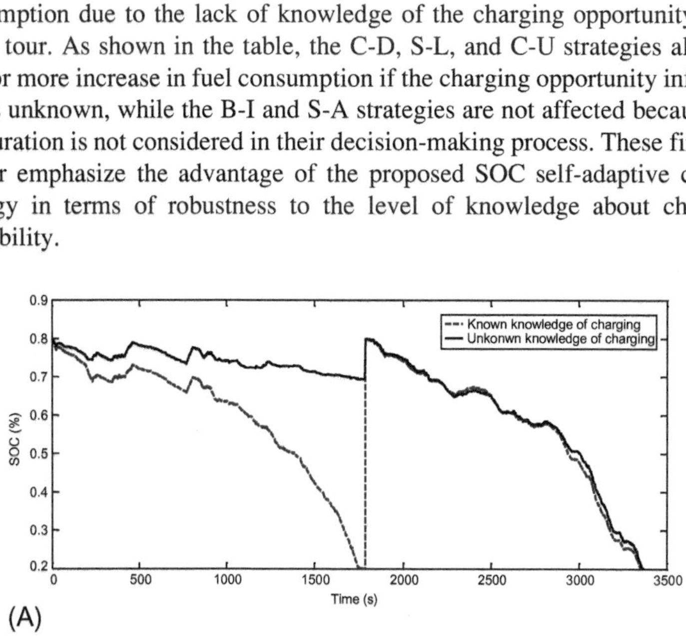

(A)

FIG. 2.16 SOC track with known or unknown charging opportunity. (A) C-D.

(Continued)

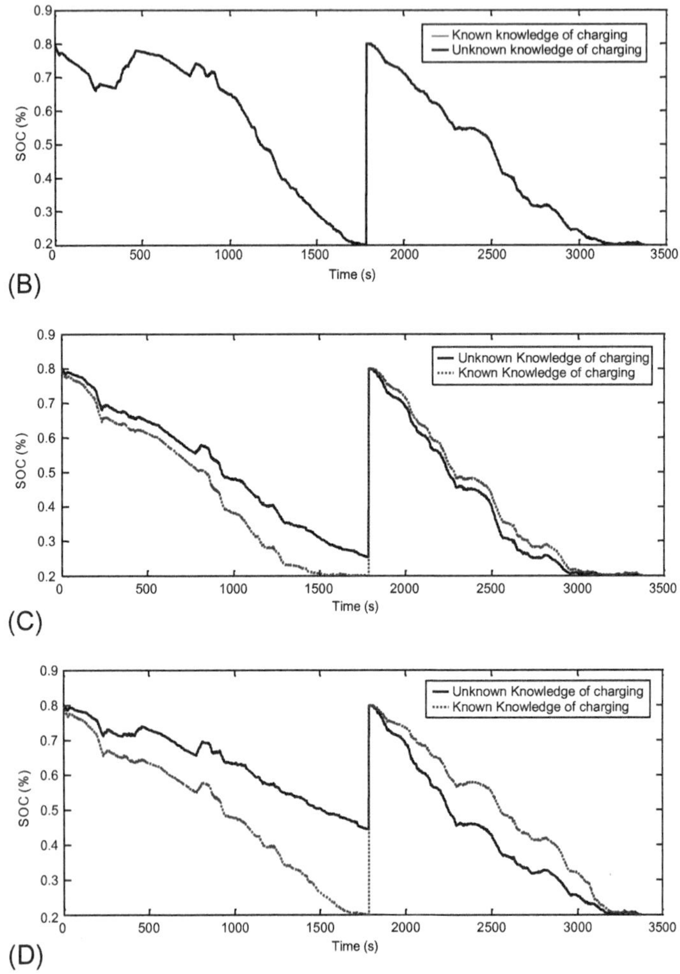

FIG. 2.16, cont'd (B) S-A. (C) C. (D) S-L.

2.5 DATA-DRIVEN REINFORCEMENT LEARNING-BASED REAL-TIME EMS

2.5.1 Introduction

As mentioned in the previous section, the EMS is at the heart of PHEV fuel economy, whose functionality is to control the power streams from both the ICE and the battery pack, based on vehicle and engine operating conditions. In the past decade, a large variety of EMS implementations have been developed for PHEVs, whose control strategies may be well categorized into two major classes, as shown in Fig. 2.17:

TABLE 2.6 Increased Fuel Consumption

Control Strategy	Known (gal)	Unknown (gal)	Increased Fuel Consumption (%)
B-I	0.9748	0.9748	00.0
B-A	0.7109	0.7543	06.1
C-D	0.6729	0.8439	25.1
S-L	0.6809	0.7853	15.0
C-U	0.7066	0.8034	13.0
S-A	0.6681	0.6681	00.0

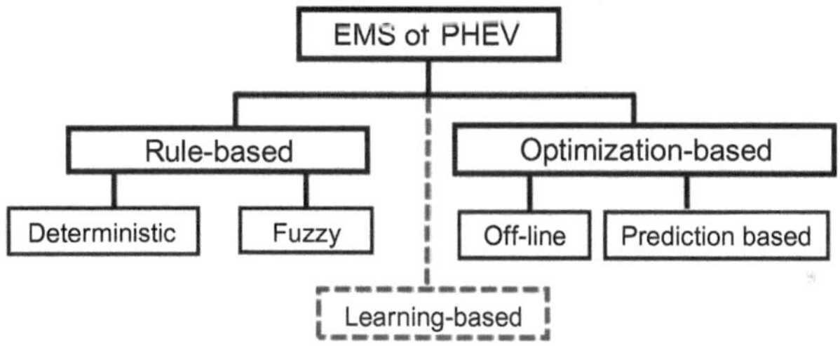

FIG. 2.17 Taxonomy of current EMS.

(a) Rule-based strategies rely on a set of simple rules without a priori knowledge of driving conditions. Such strategies make control decisions based on instant conditions only and are easily implemented, but their solutions are often far from being optimal due to the lack of consideration of variations in trip characteristics and prevailing traffic conditions; and.

(b) Optimization-based strategies are aimed at optimizing some predefined cost function according to the driving conditions and vehicle's dynamics.

The selected cost function is usually related to the fuel consumption or tailpipe emissions. Based on how the optimization is implemented, such strategies can be further divided into two groups: (1) off-line optimization which requires a full knowledge of the entire trip to achieve the global optimal solution; and (2) short-term prediction-based optimization which takes into account the predicted driving conditions in the near future and achieves local optimal solutions segment by segment within an entire trip. However, major drawbacks of these strategies include heavy dependence on the a priori knowledge of future driving conditions and high computational costs that are difficult to implement in real time.

As discussed above, there is a trade-off between the real-time performance and optimality in the energy management for PHEVs. More specifically, rule-based methods can be easily implemented in real-time but are far from being optimal, while optimization-based methods are able to achieve optimal solutions but are difficult to implement in real-time. To achieve a good balance in between, reinforcement learning (RL) has recently attracted researchers' attention. Liu et al. proposed the first and only existing RL-based EMS for PHEVs, which outperforms the rule-based controller with respect to the defined reward function but is worse in terms of fuel consumption without considering charging opportunity in the model.

In this chapter, a novel model-free RL-based real-time EMS of PHEVs is proposed and evaluated, which is capable of simultaneously controlling and learning the optimal power-split operations in real time. The proposed model is theoretically derived from DP formulations and compared to DP in the computational complexity perspective. Three major features distinguish it from existing methods:

(1) The proposed model can be implemented in real-time without any prediction efforts, since the control decisions are made only upon the current system state. The control decisions also considered for the entire trip information by learning the optimal or near-optimal control decisions from historical driving behavior. Therefore, it achieves a good balance between real-time performance and energy saving optimality.
(2) The proposed model is a data-driven model that does not need any PHEV model information once it is well trained, since all the decision variables can be observed and are not calculated using any vehicle powertrain models (these details are described in the following sections).\
(3) Compared to existing RL-based EMS implementations, the proposed strategy considers charging opportunities along the way (a key distinguishing feature of PHEVs as compared with HEVs). This proposed method represents a new class of models that could be a good supplement to the current methodology taxonomy as shown in Fig. 2.17.

2.5.2 Dynamic Programming

The above optimization problem can be solved by DP, since it satisfies Bellman's Principle of Optimality. Let s ϵ S be the state vector of the system, and a ϵ A the decision variable. The optimization problem can be converted into the following single equation given the initial state s_0 and the decisions a_t for each time step t.

$$\min_{a_t \epsilon A} E\left\{ \sum_{t=0}^{T-1} \beta^t g(s_t, s_{t+1}) | \, s_0 = s \right\} \tag{2.12}$$

where β is a discount factor and β ϵ (0,1). It can be solved by recursively calculating:

$$J(s_t) = \min_{a_t \epsilon A} E\left\{ \sum_{t=0}^{T-1} g(s_t, s_{t+1}) + \beta J(s_{t+1}) | \, s_t = s \right\}, \text{for}\, t = T-1, T-2, \ldots, 0. \tag{2.13}$$

where T is the time duration, $g(.)$ is a one-step cost function, and $J(s)$ is the true value function associated with states. Eq. (2.13) is also often referred to as Bellman's equation. In the case of PHEV energy management, s_t can be defined as a combination of vehicle states, such as the current SOC level and the remaining time to the destination, which is discussed in the following sections. a_t can be defined as the ICE power supply at each time step.

It is well known that the high computational cost of Eq. (2.13) is always the barrier that impedes its real-world application, although it is a very simple and descriptive definition. It could be computationally intractable even for a small-scale problem (in terms of state space and time span). The major reason is that the algorithm has to loop over the entire state space to evaluate the optimal decision for every single step. Another obvious drawback in the real-world application of DP is that it requires the availability of the full information of the optimization problem. In our case, it means the power demand along the entire trip should be known prior to the trip, which is always impossible in practice.

2.5.3 Approximate Dynamic Programming and Reinforcement Learning

To address the above issues, approximate dynamic programming (ADP) has been proposed (Eq. 2.13). The major contribution of ADP is that it significantly reduces the state space by introducing an approximate value function $\hat{J}(s_t, p_t)$ where p_t is a parameter vector. By replacing this approximate value function, Eq. (2.13) can be reformulated as

$$\hat{J}(s_t) = \min_{a_t \epsilon A} E\left\{ \sum_{t=0}^{T-1} g(s_t, s_{t+1}) + \beta \hat{J}(s_{t+1}, p_t) \right\}, \text{for}\, t = 0, 1, \ldots, T-1 \tag{2.14}$$

Now the optimal decision can be calculated at each time step t by

$$a_t = \arg \min_{a_t \epsilon A} E\left\{ \sum_{t=0}^{T-1} g(s_t, s_{t+1}) + \beta \hat{J}(s_{t+1}, p_t) \right\}, \tag{2.15}$$

The calculation of Eq. (2.15) now only relies on the current system state s_t, which substantially reduces the computational requirement of Eq. (2.13) to polynomial time in terms of the number of the state variables, rather than being exponential to the size of state space. In addition, the value iteration that solves the DP problem becomes forward into time, rather than being backward in Eq. (2.13). In the case of PHEV energy management, this is actually a bonus since the predicted state (e.g., power demand) at the end of the time horizon is much less reliable compared to that at the beginning of the time horizon.

In principle, the value approximate function can be learned by tuning and updating the parameter vector p_t upon the addition of each observation on state transitions. RL is an effective tool for this purpose. The specific learning technique employed in this chapter is temporal-difference learning (TD-Learning), which was originally proposed by Sutton to approximate the long-term future cost as a function of current states. The details about the implementation of the algorithm are presented in the following sections.

2.5.4 Reinforcement Learning-Based EMS

In this chapter, a TD-learning strategy is adopted for the RL problem. An action-value function: $Q\ (s,\ a)$ is defined as the expected total reward for the future receipt starting from that state. This function is to estimate "how good" it is to perform a given action in a given state in terms of the expected return. More specifically, we define $Q^{\pi}(s,a)$ as the value of taking action a in state s under a control policy π (i.e., a map that maps the optimal action to a system state), which is also the expected return starting from s, taking the action a, and thereafter following policy π:

$$Q^{\pi}(s,a) = E_{\pi}\left\{ \sum_{k=1}^{\infty} \gamma^{k} * r(s_{t+k}, a_{t+k}) \middle| s_t = s, a_t = a \right\} \tag{2.16}$$

where s_t is the state at time step t; γ is a discount factor in (0, 1) to guarantee the convergence; $r(s_{t+k}, a_{t+k})$ is the immediate reward based on the state s and action a at a given time step $(t + k)$. The ultimate goal of RL is to identify the optimal control policy that maximizes the above action-value function for all the state-action pairs.

Comparing to the formulations defined by Eqs. (2.13) and (2.14), the policy π is the ultimate decision for each time step along the entire time horizon. The reward function $r(s_{t+k}, a_{t+k})$ here is $g(.)$ in Eq. (2.13). The action-value function (i.e., $Q(s, a)$) is actually an instantiation of the approximate value function $\hat{J}(s_t)$. Thus, it is not difficult to understand that the DP formulas are the basis for a RL problem.

Conceptually, a RL system consists of two basic components: a learning agent and an environment. The learning agent interacts continuously with the environment in the following manner: at each time step, the learning agent receives an observation on the environment state. The learning agent then chooses an action that is subsequently input to the environment. The environment then moves to a new state due to the action, and the reward associated with the transition is calculated and fed back to the learning agent. Along with each state transition, the agent receives an immediate reward and these rewards are used to form a control policy that maps the current state to the best control action

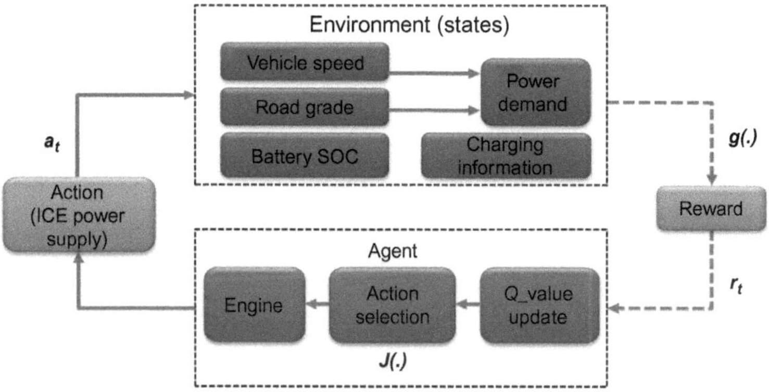

FIG. 2.18 Graphical illustration of reinforcement learning system.

upon that state. At each time step, the agent makes the decision based on its control policy. Ultimately, the optimal policy can guide the learning agent to take the best series of actions in order to maximize the cumulated reward over time that can be learned after sufficient training. Fig. 2.18 gives a graphical illustration of the learning system. The definition of the environmental states, actions, and reward are provided as follows.

2.5.5 Action and Environmental States

In this chapter, we define the discretized ICE power supply level (i.e., P_i^{eng}) as the only action the learning agent can take. The environment states include any other system parameters that could influence the decision of engine power supply. Herein we define a 5-dimensional state space for the environment, including the vehicle velocity (v_{veh}), road grade (g_{road}), percentage of remaining time to destination (t_{togo}), the battery pack's SOC (b_{soc}), and the available charging gain (c_g) of the selected charging station:

$$S=\left\{s=\left[v_{\text{veh}}, g_{\text{road}}, t_{\text{togo}}, b_{\text{soc}}, c_g\right]^T \mid v_{\text{veh}} \epsilon V_{\text{veh}}, g_{\text{road}} \epsilon G_{\text{road}}, t_{\text{togo}} \epsilon T_{\text{togo}}, b_{\text{soc}} \epsilon B_{\text{soc}}, c_c \epsilon C_g\right\}$$

where V_{veh} is the set of discretized vehicle speed level, G_{road} is the set of discretized road grade levels, and P_{brk} is the discretized level of power collected from regenerative braking (note: this power is negative compared to power demand). The minimum and maximum value of vehicle velocity, road grade, and regenerative braking power can be estimated from the historical data of commuting trips, which will be elaborated in the following section. B_{soc} is the set of battery SOC levels between the lower bound (e.g., 20%) and upper bound (e.g., 80%); T_{togo} is the percentage (10%–90%) of remaining time out of the entire trip duration, which is calculated based on the remaining distance

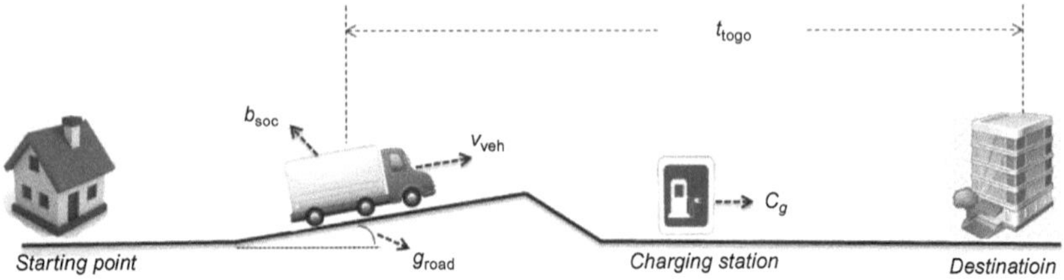

FIG. 2.19 Illustration of environment states along a trip.

to destination. C_g is the set of discretized charging gain (e.g., 30%, 60%) of the selected charger. This charging gain represents the availability of the charger, which may be a function of the charging time and charging rate, and is assumed to be known beforehand. It is noteworthy that all the states can be measured and updated in real time as the vehicle is running. Fig. 2.19 shows all the real-time environmental states.

2.5.6 Reward Initialization (With Optimal Results From Simulation)

The definition of reward is dependent upon the control objective, which is to minimize the fuel cost while satisfying the power demand requirement. Hence, we define the reciprocal of the resultant ICE power consumption at each time step as the immediate reward. A penalty term is also included to penalize the situation where the SOC is beyond the predefined SOC boundaries. The immediate reward is calculated by the following equations:

$$r_{ss'}^{a} = \begin{cases} \dfrac{1}{P_{\text{ICE}}} \text{ if } P_{\text{ICE}} \neq 0 \text{ and } 0.2 \leq \text{SOC} \leq 0.8 \\ \dfrac{1}{P_{\text{ICE}} + P} \text{ if } P_{\text{ICE}} \neq 0 \text{ and} (\text{SOC} \leq 0.2 \text{ or } \text{SOC} \geq 0.8) \\ \dfrac{2}{Min_{P_{\text{ICE}}}} \text{ if } P_{\text{ICE}} = 0 \text{ and } 0.2 \leq \text{SOC} \leq 0.8 \\ \dfrac{1}{2*P} \text{ if } P_{\text{ICE}} = 0 \text{ and } (\text{SOC} \leq 0.2 \text{ or } \text{SOC} \geq 0.8) \end{cases} \tag{2.17}$$

where $r_{ss'}{}^{a}$ is the immediate reward when state changes from s to s' by taking action a, P_{ICE} is the ICE power supply, P is the penalty value and is set as the maximum power supply from ICE in this chapter, and Min_P_{ICE} is the minimum nonzero value of ICE power supply. This definition guarantees that the minimum ICE power supply (action) which satisfies the power demand as well as SOC constraints can have the largest numerical reward. A good initialization of reward is also critical for the quick convergence of the proposed algorithm. In this case, the optimal or near-optimal results of typical trips obtained from simulation are used as the initial seeds. These optimal or near-optimal results are deemed as the control decisions made by "good drivers" from historical driving. In order to obtain a large number of such good results for algorithm training, an EA is adopted for the off-line full-trip optimization, since EA can provide multiple solutions for one single run. These solutions are of different quality, which may well represent different levels of driving proficiency in a real-world situation.

2.5.7 Q-Value Update and Action Selection

In the algorithm, a Q value, denoted by $Q(s, a)$, is associated with each possible state-action pair (s, a). Hence there is a Q table which is updated during the learning process and can be interpreted as the optimal control policy that the learning agent is trying to learn. At each time step, the action is selected upon this table after it is updated. The details of the algorithmic process are given in the following pseudo code:

Algorithm: RL based PHEV EMS algorithm

Inputs: Initialization 6-D $Q(s, a)$ table; discount factor $\gamma = 0.5$; learning rate $\alpha = 0.5$; exploration probability $\varepsilon \in (0,1)$; vehicle power demand profile P_d (N time steps)

Outputs: $Q(s, a)$ array; control decisions P_d (T time steps)

```
Initialize Q (s, a) arbitrarily
For each time step t = 1: T
    Observe current s_t (v_veh, g_road, t_togo, b_soc, C_g)
    Choose action a_t for the current state s_t:
            temp = random (0,1);
            if temp <= ε
                a_t = arg max_{a∈A} { Q(s_t, a)}
            else
                a_t = randomly choose an action;
            end
        Take action a_t, observe next state s_{t+1} (P_{t+1}, SOC_{t+1})
    if SOC_{t+1} < 0.2
        Switch into charging-sustaining mode;
        Give big penalty to r_t according to Eq. (2.10)
    else
        Calculate reward r_t according to Eq. (2.10)
    end
    Update Q (s_t, a_t) with following value:
      Q(s_t, a_t) + α{r_t + γ* max_{a_{t+1}} {Q(s_{t+1}, a_{t+1})} − Q(s_t, a_t)}
end
```

2.5.8 Validation and Testing

The proposed model is then evaluated with real-world data in two different scenarios: one without charging opportunities and the other with charging opportunities.

To obtain a series of real trip data (second-by-second velocity trajectories), we apply the trajectory synthesis technique proposed in our previous work to the inductive loops detector (ILD) data archived in the California Freeway Performance Measurement System (PeMS).

The trajectory synthesis is a two-step process: (1) estimating average velocity by applying 2-dimensional interpolation method to real-world traffic data (e.g., volumes and occupancy) collected from ILDs; and (2) generating random velocity disturbance based on representative driving cycles from the MOVES (MOtor Vehicle Emission Simulator) database. Real traffic data were collected at the I-210 freeway segment between I-605 and Day Creek Blvd in southern California, starting at 8:00 a.m. in the morning (westbound) and returning at 4:00 p.m. in the afternoon every weekday during the period between January 9, 2012 and January 17, 2012. Twelve trips (including eastbound and westbound) were generated in total. The road grade information was also synchronized with the trip data to estimate the second-by-second power demands. For more detailed information on the trajectory synthesis and power demand profile generation, please refer to [15].

2.5.9 Model Without Charging Opportunity (Trip Level)

To validate the proposed strategy, the model without considering charging opportunity is first trained and tested with trips where there is no charging opportunity within the trip. Data for multiple westbound trips described in [15] are used for training. Although it has been proven that Q-learning is guaranteed to converge mathematically, an experimental analysis of convergence is conducted in this chapter. In the experiment, the trip data for the first 6 days are concatenated one by one to form a single training cycle. The proposed model is trained with repeated training cycles. At the end of each training cycle, the trained model is tested with the 7th day trip, and the fuel consumption is recorded in Fig. 2.20. In addition, the training with or without good initialization using simulated optimal or near-optimal solution are also compared. As we can see in the figure, there is a clear convergence in fuel consumption for both cases. However, the initialization with simulated optimal or near-optimal solutions help achieve a faster convergence.

As previously described, the selected state space is 5-dimensional and the action space has 1 dimension. Therefore the Q (s, a) table is 6-dimensional. Fig. 2.21 shows the 4-D slice diagram of the learned Q (s, a) table in which different color grids represent different numerical reward values (e.g., darker color means lower values) and three slices on the (ICE power supply, power demand)

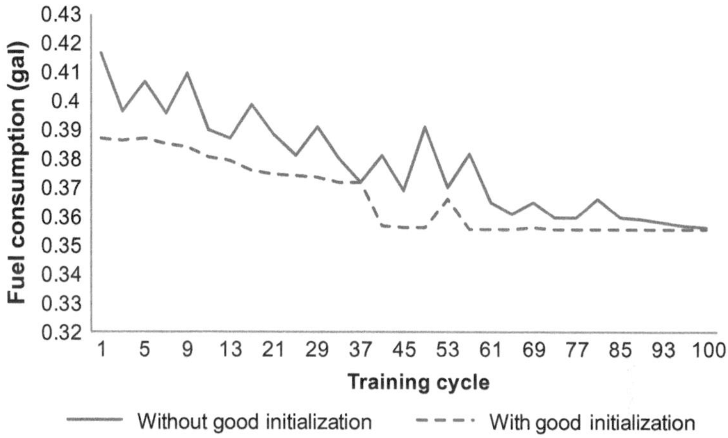

FIG. 2.20 Convergence analysis ($\varepsilon = 0.7$; $\gamma = 0.5$; $\alpha = 0.5$).

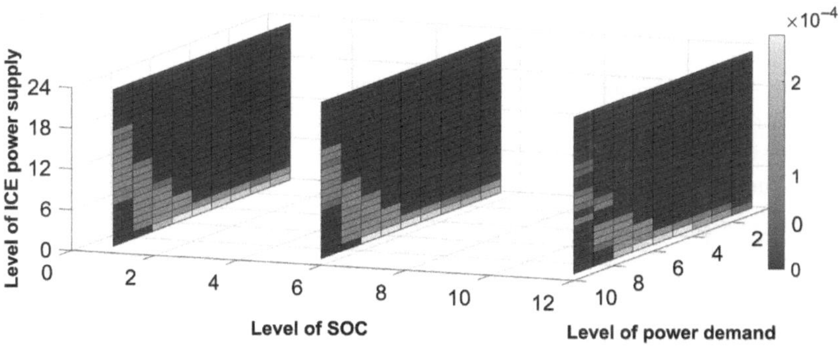

FIG. 2.21 4-D slice diagram of the learned Q table.

space are given at three different SOC levels: 1, 6, and 12 (i.e., 20%, 50%, and 80%). Please note that the road grade and vehicle speed are implicitly aggregated into power demand. The dimension of remaining time is not indicated in the figure. As can be observed in each slice, when the power demand is not so high (e.g., below level 5), action level 1 or 2 is usually the most appropriate because the least ICE power is consumed. When the power demand becomes higher, the range of the feasible action levels gets wider also. In such cases, lower levels of ICE power supply may not be enough to satisfy the power demand and the resultant SOC level could be lower than 0.2, resulting in a penalty defined in Eq. (2.17). It is also noted that when SOC level is high, it is less likely that the higher ICE power supply level would be chosen to satisfy the same power demand. This is because when the vehicle battery SOC is high, the ICE power is not likely to be used aggressively.

As discussed in the previous sections, an exploration-exploitation strategy is adopted for the action selection process to avoid premature convergence. The action with the biggest Q value has a probability of $1 - \varepsilon$ to be selected. Hence the value of ε may significantly affect the performance of the proposed method. To evaluate such impacts, a sensitively analysis of ε is carried out and illustrated in Fig. 2.8. It can be observed that both the fuel consumption and the resultant SOC curves are very close to those of the binary mode control if the value of ε is small. A possible explanation is that a small ε value indicates a large probability to select the most aggressive action with the biggest Q value (or the lowest levels of ICE power supply). Therefore, the battery power is consumed drastically as it is with the binary mode control. However, if the value of ε is too large (e.g., >0.8), the battery power is utilized too conservatively where the final SOC is far away from the lower bound, resulting in much greater fuel consumption. It is found that the best value of ε in this chapter is around 0.7, which ensures the SOC curve is quite close to the global optimal solution obtained by the off-line DP strategy. With this best ε value, the fuel consumption is 0.3559 gal, which is 11.9% less than that of the binary mode control and only 2.86% more than that of DP strategy as shown in Fig. 2.8. This also implies that an adaptive strategy for determining exploration rate along the trip could be useful. Fig. 2.9A shows a linearly decreasing control of ε along the trip. A smaller ε is preferred at the later stage of the trip because SOC is low and the battery power should be consumed more conservatively. With this adaptive strategy for ε, the proposed mode could also achieve a good solution with 0.3570 gal of fuel consumption, which is 11.7% less than the binary control shown in Fig. 2.22.

2.5.10 Model With Charging Opportunity (Tour Level)

The most distinctive characteristics of PHEVs from HEVs is that PHEV can be externally charged whenever a charging opportunity is available. To evaluate further the impacts due to charging availability, we include this information

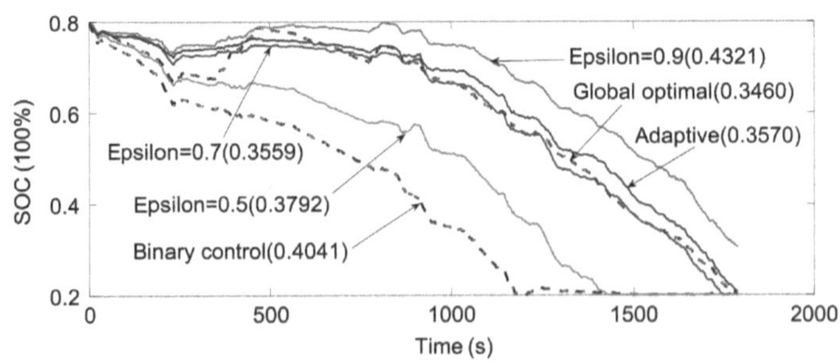

FIG. 2.22 Fuel consumption in gallon *(bracketed values)* and SOC curves by different exploration probabilities.

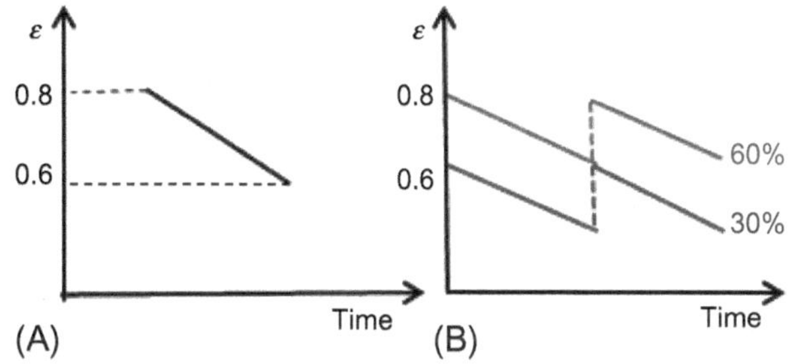

FIG. 2.23 (A) Linear adaptive control of ε; (B) linear adaptive control of ε with charging opportunity.

in the proposed model as a decision variable. For simplicity, the charging opportunity is quantified by the gain in the battery's SOC, which may be a function of available charging time and charging rate. Data for a typical tour are constructed by combining a round trip between the origin and destination. We assume there is a charger in the working place (west-most point in the map) and the available charging gain has only two levels: 30%–60%. In this case, a corresponding adaptive strategy of ε is also used as shown in Fig. 2.23B. The rationale behind this adaptive strategy is that battery power should be used less conservatively (i.e., higher ε value) after charging, and/or when C_g is higher.

The obtained optimal results are shown in Figs. 2.24 and 2.25. As we can see in both figures, the resultant SOC curves are much closer to the global optimal solutions obtained by DP than the binary control. To obtain a statistical significance of the performance, the proposed model is tested with 30 different trips

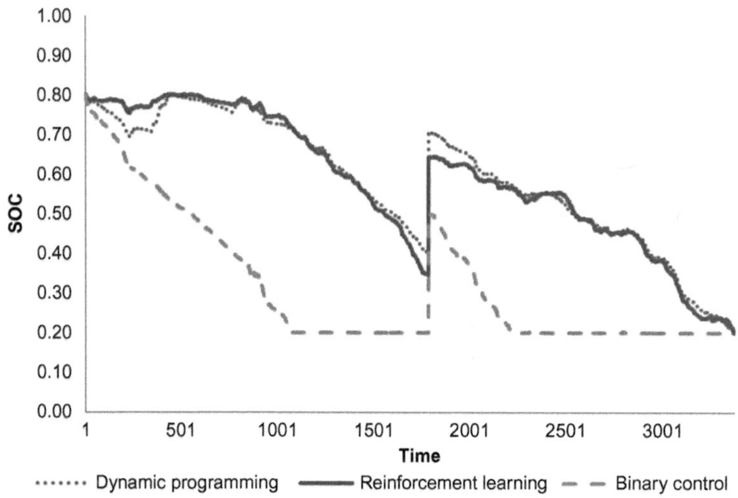

FIG. 2.24 Optimal results when available charging gain is 0.3 ($C_g = 0.3$).

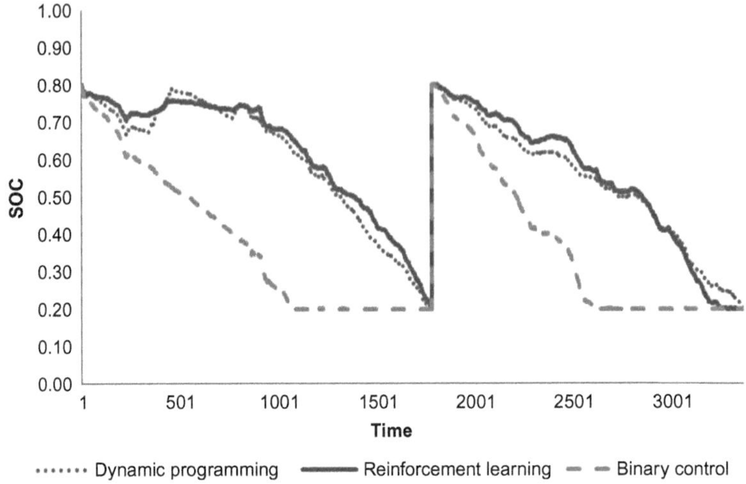

FIG. 2.25 Optimal results when available charging gain is 0.6 ($C_g = 0.6$).

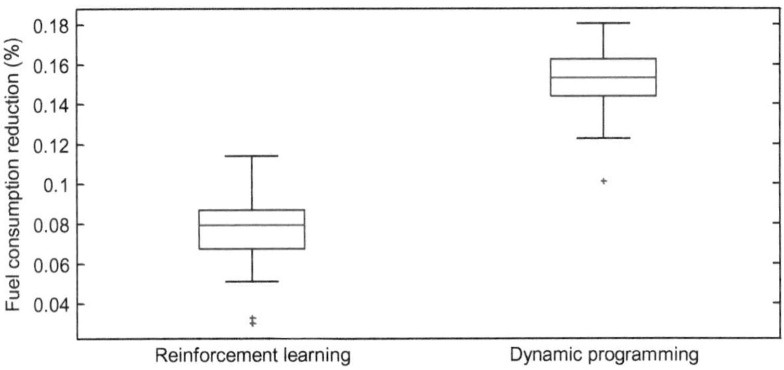

FIG. 2.26 Fuel consumption reduction compared to binary control.

by randomly combining two trips and a charging station is assumed in between with a random C_g (randomly choose from 30% to 60%). By taking the binary control as baseline, the reduced fuel consumption is given in Fig. 2.26. As we can see in the figure, RL model achieves an average of 7.9% fuel savings. It seems that having more information results in lower fuel savings, which is counterintuitive. The reason is that the inclusion of additional information or state variable to the model exponentially increases the search space of the problem, which thereby increases the difficulty of learning the optimal solution. In addition, more uncertainty is introduced to the learning process due to the errors within the added information, which degrades the quality of the best solution the model can achieve.

2.6 CONCLUSIONS

In this chapter, two different data-driven on-line EMS for PHEVs, (i.e., EA based EMS and RL based EMS) are presented with both system architecture and real-world validation.

For the EA-based EMS, the proposed framework applies the self-adaptive strategy to the control of the vehicle's SOC in a rolling horizon manner for the purpose of real-time implementation. The control of the vehicle's SOC is formulated as a combinatory optimization problem that can be efficiently solved by the EDA. The proposed EMS is evaluated comprehensively using a number of trip profiles extracted from real-world traffic data. The results show that the self-adaptive control strategy used in the proposed system statistically outperforms the conventional binary control strategy with an average of 10.7% fuel savings. The sensitivity analysis reveals that the optimal prediction horizon window of the proposed EMS is 250 s, which requires 5.8 s of computation time in our study case. This amount of time is much less than the optimal control horizon window of 10 s, which confirms the feasibility of real-time implementation. Another important advantage of the proposed EMS is that, unlike other existing systems, it does not require a priori knowledge about the trip duration. This allows the proposed system to be robust against real-world uncertainties, such as unexpected traffic congestion that increases the trip duration significantly, and changes in intertrip charging availability.

For RL-based EMS, it is capable of simultaneously controlling and learning the optimal power-split operation. The proposed EMS model is tested with trip data (i.e., multiple speed profiles) synthesized from real-world traffic measurements. Numerical analyses show that a near-optimal solution can be obtained in real-time when the model is well trained with historical driving cycles. For the study cases, the proposed EMS model can achieve better fuel economy than the binary mode strategy by about 12%–8% at the trip level and tour level (with charging opportunity), respectively. The possible topics for future work are: (1) propose a self-adaptive tuning strategy for exploration-exploitation (ε); (2) test the proposed model with more real-world trip data which could include other environmental states, such as the position of charging stations; and (3) conduct a robustness analysis to evaluate the performance of the proposed EMS model when there is error present in the measurement of environment states.

REFERENCES

[1] U.S. Department of Transportation, http://www.its.dot.gov/research/vehicle_electrification_smartgrid.htm. (Accessed 5 January 2015).

[2] G. Wu, K. Boriboonsomsin, M. Barth, Development and evaluation of an intelligent energy-management strategy for plug-in hybrid electric vehicles, IEEE Trans. Intell. Transp. Syst. 15 (3) (2014) 1091–1100.

[3] S.G. Wirasingha, A. Emadi, Classification and review of control strategies for plug-in hybrid electric vehicles, IEEE Trans. Veh. Technol. 60 (1) (2011) 111–122.

[4] A. Panday, H.O. Bansal, A review of optimal energy management strategies for hybrid electric vehicle, Int. J. Veh. Technol. (2014) 19.

[5] H. Banvait, S. Sohel, Y. Chen, in: A rule-based energy management strategy for plug-in hybrid electric vehicle (PHEV), Proceedings of American Control Conference, St. Louis, MO, June 2009, 2009, pp. 3938–3943.

[6] Q. Gong, Y. Li, in: Trip based optimal power management of plug-in hybrid electric vehicles using gas-kinetic traffic flow model, Proceedings of American Control Conference, Seattle, WA, June 2008, 2008, pp. 3225–3230.

[7] L. Tribiloli, M. Barbielri, R. Capata, E. Sciubba, E. Jannelli, G. Bella, A real time energy management strategy for plug-in hybrid electric vehicles based on optimal control theory, Energy Procedia 45 (2014) 949–958.

[8] C. Hou, L. Xu, H. Wang, M. Ouyang, H. Peng, Energy management of plug-in hybrid electric vehicles with unknown trip length, J. Frankl. Inst. 352 (2) (2015) 500–518.

[9] M. Vajedi, M. Chehrehsaz, N.L. Azad, Intelligent power management of plug-in hybrid electric vehicles, part I: real-time optimum SOC trajectory builder, Int. J. Electr. Hybrid Veh. 6 (1) (2014) 46–67.

[10] N. Denis, M.R. Dubois, A. Desrochers, Fuzzy-based blended control for the energy management of a parallel plug-in hybrid electric vehicle, IET Intell. Transp. Syst. 9 (1) (2015) 30–37.

[11] X. Wang, H. He, F. Sun, X. Sun, H. Tang, Comparative study on different energy management strategies for plug-in hybrid electric vehicles, Energies 6 (2013) 5656–5675.

[12] W. Jian, in: Fuzzy energy management strategy for plug-in hev based on driving cycle modeling, Control Conference (CCC), 2014 33rd Chinese, 28–30 July, 2014, pp. 4472–4476.

[13] L. Tribioli, S. Onori, in: Analysis of energy management strategies in plug-in hybrid electric vehicles: application to the GM Chevrolet volt, American Control Conference (ACC), 17–19 June 2013, 2013, pp. 5966–5971.

[14] H. Yu, M. Kuang, R. McGee, Trip-oriented energy management control strategy for plug-in hybrid electric vehicles, IEEE Trans. Control Syst. Technol. 22 (4) (2014) 1323–1336.

[15] Q. Gong, Y. Li, Z.-R. Peng, in: Trip based optimal power management of plug-in hybrid electric vehicles using gas-kinetic traffic flow model, American Control Conference, 11–13 June 2008, 2008, pp. 3225–3230.

[16] M.P. O'Keefe, T. Markel, Dynamic Programming Applied to Investigate Energy Management Strategies for a Plug-In HEV, National Renewable Energy Laboratory, Golden, CO, 2006. Report No. NREL/CP-540-40376.

[17] Z. Chen, C.C. Mi, R. Xiong, X. Jun, C. You, Energy management of a power-split plug-in hybrid electric vehicle based on genetic algorithm and quadratic programming, J. Power Sources 248 (15) (2014) 416–426.

[18] T. Feng, L. Yang, Q. Gu, Y. Hu, T. Yan, B. Yan, A supervisory control strategy for plug-in hybrid electric vehicles based on energy demand prediction and route preview, IEEE Trans. Veh. Technol. PP (99) (2015) 1.

[19] V. Larsson, L. Johannesson Mårdh, B. Egardt, S. Karlsson, Commuter route optimized energy management of hybrid electric vehicles, IEEE Trans. Intell. Transp. Syst. 15 (3) (2014) 1145–1154.

[20] C. Sun, S.J. Moura, X. Hu, J.K. Hedrick, F. Sun, Dynamic traffic feedback data enabled energy management in plug-in hybrid electric vehicles, IEEE Trans. Control Syst. Technol. 23 (3) (2015) 1075–1086.

[21] M. Vajedi, A. Taghavipour, N.L. Azad, J. McPhee, in: A comparative analysis of route-based power management strategies for real-time application in plug-in hybrid electric vehicles, American Control Conference (ACC), 2014, Portland, OR, 2014, pp. 2612–2617.

[22] Z. Chen, W. Liu, Y. Yang, W. Chen, Online energy management of plug-in hybrid electric vehicles for prolongation of all-electric range based on dynamic programming, Math. Probl. Eng. 2015 (2015). 11 pages.

[23] S.J. Moura, H.K. Fathy, D.S. Callaway, J.L. Stein, A stochastic optimal control approach for power management in plug-in hybrid electric vehicles, IEEE Trans. Control Syst. Technol. 19 (3) (2011) 545–555.

[24] C. Liu, Y.L. Murphey, in: Power management for plug-in hybrid electric vehicles using reinforcement learning with trip information, Transportation Electrification Conference and Expo (ITEC), 2014 IEEE, 15–18 June, 2014, pp. 1–6.

[25] X. Lin, H. Banvait, S. Anwar, Y. Chen, in: Optimal energy management for a plug-in hybrid electric vehicle: real-time controller, American Control Conference (ACC), June 30, 2010–July 2, 2010, 2010, pp. 5037–5042.

[26] X. Qi, G. Wu, K. Boriboonsomsin, M.J. Barth, J. Gonder, Data-driven reinforcement learning-based real-time energy management system for plug-in hybrid electric vehicles, Transp. Res. Rec.: J. Transp. Res. Board 2572 (2016) 1, https://doi.org/10.3141/2572-01.

[27] M. Hauschile, M. Pelican, An Introduction and Survey of Estimation of Distribution Algorithms, MEDAL Report No. 2011004, University of Missouri-St. Louis, 2011.

[28] X. Qi, K. Rasheed, K. Li, W.D. Potter, in: A fast parameter setting strategy for particle swarm optimization and its application in urban water distribution network optimal design, The 2013 Int'l. Conf. on Genetic and Evolutionary Methods (GEM), 2013.

[29] X. Qi, Swmarm Intelligence Inspired Engineering Optimization: Concepts, Modeling and Evaluation, Lambert Academic Publishing House, 2014. ISBN: 978-3-659-35681-0.

[30] X. Qi, G. Wu, K. Boriboonsomsin, M.J. Barth, in: An on-line energy management strategy for plug-in hybrid electric vehicles using an estimation distribution algorithm, IEEE 17th International Conference on Intelligent Transportation Systems (ITSC), 2014, pp. 2480–2485. 8–11 October 2014.

[31] A.E. Eiben, Introduction to Evolutionary Computing, Springer, 2007.

[32] P.S. Oliveto, J. He, X. Yao, Time complexity of evolutionary algorithms for combinatorial optimization: a decade of results, Int. J. Autom. Comput. 4 (2007) 281–293.

[33] D. Kum, Modeling and Optimal Control of Parallel HEVs and Plug-In HEVs for Multiple Objectives (Ph.D. Dissertation), University of Michigan, 2010.

Chapter 3

Machine Learning and Computer Vision-Enabled Traffic Sensing Data Analysis and Quality Enhancement

Guohui Zhang* and Yinhai Wang†

*Department of Civil and Environmental Engineering, University of Hawaii at Manoa, Honolulu, HI, United States, †Department of Civil and Environmental Engineering, University of Washington, Seattle, WA, United States

Chapter Outline

3.1 INTRODUCTION

3.1.1 Significance of Vehicle Classification Volumes

Due to the considerable difference in characteristics between long vehicles (LVs) and short vehicles (SVs), accurate and timely vehicle classification data

Data-Driven Solutions to Transportation Problems. https://doi.org/10.1016/B978-0-12-817026-7.00003-5

are of fundamental importance. Various transportation issues including economy, efficiency, environment, safety, and enforcement are associated with traffic volume patterns. For example, large trucks, buses, and recreational vehicles are typically associated with slow acceleration, inferior braking, and a large turning radius. Previous studies have found that the percentage of trucks in a traffic stream has significant effects on traffic capacity and safety. The Highway Capacity Manual 2010 [1] requires adjustments to the volumes of these vehicles in highway design and capacity analysis. The geometric design of a roadway, such as horizontal alignment and curb heights, is affected by the different moving characteristics of LVs due to their heavy weight, inferior braking, and large turning radius. Due to the heavy weight of these vehicles when they are fully loaded, their impact to pavement lifetime is significant, and the volumes of these vehicles are indispensable inputs for pavement design and maintenance [2]. From a safety perspective, these large vehicle volumes are also desirable because statistics show that large vehicles have a higher risk of accident, and traffic accidents with these vehicles involved are more severe. For example, although large trucks are only 4% of total registered vehicles in the United States, they account for 8% of all vehicles involved in fatal crashes [3]. Recent studies [4,5] also found that particulate matters (PM) are strongly associated with the onset of myocardial infarction and respiratory symptoms. Heavy-duty trucks that use diesel engines are major sources of PM, accounting for 72% of traffic emitted PM [6]. Therefore, classified vehicle volumes are important for traffic operation, pavement design, and transportation planning.

3.1.2 Research Motivation

All these facts illustrate that truck volume data are extremely important for accurate analysis of traffic safety, traffic pollution, and flow characteristics in transportation planning, management, and engineering. Unfortunately, most traffic sensors such as single-loop detectors currently in place cannot directly measure truck volumes. Although dual-loop detectors provide classified vehicle volumes, there are too few of them on our current freeway systems to meet practical needs. Upgrading a single-loop detector to a dual-loop detector requires putting in another single-loop to pair up with the existing single-loop. The cost of such an upgrade is expensive considering the hardware cost and the indirect cost resulting from the necessary lane closure. Other traffic detectors, such as the Triple-Technology traffic detector (TT-298), etc., are capable of measuring classified vehicle volumes, but are not widely available in the existing roadway infrastructure. Single-loop detectors and traffic surveillance cameras have been increasingly deployed for monitoring traffic status on major roadways. Effective utilization of these single-loop detectors and cameras for traffic data collection is desirable considering their safer and less costly installation and maintenance. On the other hand, machine learning and computer vision techniques have been developed rapidly in recent years. As two major artificial intelligence and advanced computing techniques, machine learning, and computer vision can substantially revolutionize existing traffic sensing practices and theoretical foundations. In this study, we are motivated to develop a machine learning approach to establish an ANN to better extract classified vehicle volumes from single-loop measurements. In addition, a set of computer vision-based algorithms are developed to calculate vehicle lengths for classification based on widely available surveillance camera signals.

3.1.3 Research Objectives

The research objectives of this study include the following:

(1) Based on machine learning principles, develop an ANN method for vehicle bin-volume estimation using single-loop data.
(2) Develop a series of computer vision-based algorithm to measure accurate vehicle detection and classification data without the need for complicated calibration using surveillance camera signals.
(3) Conduct experimental tests under various traffic operation scenarios to verify the effectiveness of the developed ANN approach and the image processing algorithms for vehicle classification volume data collection.

This chapter is organized as follows. Before presenting the details of the ANN method and image processing algorithms in the methodology section, related studies are briefly introduced. Experimental design and tests are conducted for these two detection approaches, and their performance is discussed in the section following the methodology. The final section concludes this research effort and proposes further research topics.

3.2 STATE OF THE ART AND PRACTICE

3.2.1 Single-Loop Vehicle Length Estimation and Machine Learning Application

A single-loop detector outputs vehicle volume and lane occupancy periodically. For example, the California Department of Transportation (Caltrans) loop detectors output data every 30 s. Many algorithms have been developed to estimate traffic classification volumes from single-loop outputs. Several of these algorithms were based on Athol's speed-estimation formula [7]. These studies focused on improving speed-estimation accuracy by choosing the right length estimation parameter (commonly referred to as the g factor) or preprocessing single-loop data before applied to Athol's speed-estimation formula. Coifman et al. [8] recommended using the median lane-occupancy per vehicle for speed estimation. Wang and Nihan [9] suggested a filtering process to screen out intervals with LVs and using only short-vehicle measurements for speed calculation. Coifman [10] proposed calibrating g in free flow condition when traffic speed is known and applying the calibrated g for speed estimation in other time periods.

Many researchers, such as Mikhalkin [11], have sought sophisticated filtering methods to improve vehicle length and speed estimates based on data mining and machine learning techniques. Dailey [12] considered random errors in the measurements and used a Kalman filter for speed estimation. Pushkar et al. [13] developed a cusp catastrophe theory model to estimate speed. Sun and Ritchie [14] proposed a linear model to estimate individual vehicle speeds using slew rates of single-loop inductive waveforms. They concluded that their proposed algorithm performed better than conventional speed-estimation methods. The estimated speed enables vehicle-length calculation and length-based vehicle classification. Kwon [15] developed an algorithm to estimated truck volumes in multilane freeway using lane-to-lane speed correlation. Mittal [16] proposed a statistical approach to estimate truck volumes on state highways. Sun et al. [14] used waveforms to extract vehicle lengths for vehicle reidentification. Wang and Nihan [9] proposed a vehicle classification method based on the nearest neighbor decision rule for classifying vehicles into two categories (short vehicle (SV) and long vehicle (LV)) using single-loop measurements. While this algorithm produced reasonably accurate vehicle classification, the classification categories were rough. It would be better to classify vehicles into the four categories as has been done by the WSDOT dual-loop detectors. To reach such a goal, new research efforts are required.

3.2.2 Computer Vision-Based Traffic Detection

Applying image processing technologies to vehicle detection has been a popular focus of research in Intelligent Transportation Systems (ITS) over the last decade. The early video detection research [17] at the University of Minnesota has resulted in the Autoscope video detection systems that are widely used in today's traffic detections and surveillance around the world. Lai et al. [18] demonstrated that accurate vehicle dimension estimation could be performed using a set of coordinate mapping functions. Similarly, commercially available Video Image Processors (VIPs), such as the VideoTrack system developed by Peek Traffic Inc., are capable of truck data collection. Calibrating these systems normally requires very specific road surface information (such as the distance between recognizable road surface marks) and camera information (such as the elevation and tilt angle), which may not be easy to obtain [19]. Furthermore, some studies [20–22] evaluating some of these commercial systems found that shadows and headlight reflections generated significant problems of false detections and early detections. Hasegawa and Kanade [23] developed a system capable of detecting and classifying moving objects by both type and color. Vehicles from a series of training images were identified by an operator to develop the characteristics associated with each object type. In a test of 180 presented objects, 91% were correctly identified. Rad and Jamzad [24] developed a program to count and classify vehicles as well as identify the occurrence of lane-changes through tracking. Their approach utilized a background subtraction approach combined with morphological operations to identify moving vehicle regions. Graettinger et al. [25] used video data collected from an Autoscope Solo Pro commercial detection system to provide classifications corresponding to the 13 FHWA vehicle classes. The method was tested at one location and validated at four other sites. However, use of site-specific models is less feasible since new models for each location would have to be developed. These previous studies provided valuable insights to the video-based vehicle detection and classification (VVDC) problems to be addressed in this study. However, several vital problems impede the wide applications of these existing algorithms, such as complicated calibration processes, vehicle occlusions, camera vibrations, weather and light conditions, and shadow impacts. Although some considerations were proposed to address parts of these problems, the performance of the integrated systems is not as good as expected. They were mostly constrained by service conditions in practicality. Therefore, we tend to develop a new video system for vehicle detection and classification by incorporating new robust algorithms.

3.3 METHODOLOGY

3.3.1 Machine Learning Approach for Vehicle Classification Volume Estimation

3.3.1.1 Artificial Neural Network (ANN) Overview

Due to the stochastic features of traffic flow, deterministic mathematical equations based on certain assumptions for speed calculation typically do not work well for all situations and may result in significant speed-estimation errors under certain traffic conditions. When estimated speed is used in vehicle-length calculation, the secondary estimation procedure will generate significant errors cumulated from each estimation steps. Such problems, however, cannot be overcome by deterministic mathematical equations. ANN appears to be an effective solution to such a problem. ANN is a powerful data modeling tool that is able to capture and represent complex input/output relationships and characteristics, such as associativity, self-organization, generalizability, and noise- and fault-tolerance [26,27]. Along with the development of computing science, modern information processing technologies, such as genetic algorithm, expert systems, etc., and ANN technologies have developed quickly. ANN has been extensively used in many transportation studies and has proven an effective solution to problems that are too complicated to be represented and optimized by conventional mathematical methods [28–31]. Therefore, we propose to use ANN for capturing the complicated relationships between single-loop measured variables and classified vehicle volumes under various traffic conditions.

3.3.1.2 ANN Architecture and Algorithm

Based on the observed traffic characteristics and the efforts of trial and error, we design a three-layer, feed-forward ANN with the architecture of back-propagation (BP), one of the most popular and stable network architectures, to estimate classified vehicle volumes from single-loop measurements. Standard BP is a gradient descent algorithm; for our problem we adopt the Levenberg-Marquardt algorithm as the training rule, in which the network weights are moved along the negative of the gradient of the performance function. The term BP refers to the manner in which the gradient is computed for nonlinear multilayer networks. Properly trained BP networks tend to give reasonable answers when presented with inputs that they have never seen. Typically, a new input similar to the input vectors used in training can lead to an output close enough to the correct output. This generalization property makes it possible to train a network with a representative set of input pairs and get good results without training the network using all possible input and output pairs. Following this idea, vehicle classification can be conducted independently without applying the estimated speed data. This kind of straightforward estimation avoids accumulating errors resulting from inaccurate speed estimates. For this study, vehicles are divided into four classes that are consistent with the four bins used by the Washington State Department of Transportation (WSDOT) detection system. The four length-based vehicle categories are described in Table 3.1.

In order to mine more associated relationships between a series of single-loop measurements and the corresponding bin volumes, we employ multidimensional input vectors to train the proposed ANN. In addition, different structures are adopted to best fit the specific properties of each vehicle category. For instance, the neural network for estimating Bin 1 volume is designed to have 19 nodes in the input layer: one node for the time stamp input and nine pairs of nodes for inputting single-loop measurements (volume and lane occupancy) over a 3-min period (there are nine 20-s intervals in a 3-min period). As a rule, a network with too few hidden units only occasionally discovers hidden dependencies in data sets, and the network is likely to produce a significant number of errors. On the other hand, a network with too many hidden units tends to

TABLE 3.1 Four Length-Based Vehicle Categories Used by the WSDOT

Classes	Range of Length	Vehicle Types
Bin 1	Less than 26 ft (7.92 m)	Cars, pickups, and short single-unit trucks
Bin 2	From 26 ft (7.92 m) to 39 ft (11.89 m)	Cars and trucks pulling trailers, long single-unit trucks
Bin 3	From 40 ft (12.19 m) to 65 ft (19.81 m)	Combination trucks
Bin 4	Longer than 65 ft (19.81 m)	Multitrailer trucks

memorize all data instead of finding the associated relations and this often leads to an ineffective model with remarkable network errors. The hidden layer of this study is designed to have 35 nodes in order to provide the capacity for approximating and converging with the proposed ANN and to balance other factors of the sample data set. The output layer contains one node for Bin 1 volume output. The network structure for Bin 1 volume estimation can be briefly expressed as 19-35-1. Analogously, similar design procedures have been conducted for other vehicle categories. For each bin category, a different network structure distinguished by the number of nodes on the hidden layer is applied to discover and store the implicit interrelationships among single-loop measurements and the bin volume. Their input vectors are the same as that of Bin 1, which consists of one time stamp, nine continuous 20-s-sampling volumes and nine corresponding occupancies in 3 min. The network structures for estimating the volumes of Bin 2, Bin 3, and Bin 4 are designed as 19-8-1, 19-5-1, and 19-21-1, respectively, based on the characteristics of the data and results of trial and error. The entire network architecture is shown in Fig. 3.1.

Though different network structures are proposed for estimating volumes of different vehicle categories, the same calculation procedure is adopted and implemented for all four networks. This calculation procedure is described as follows. Every element of the input vector is assigned an independent weighting factor w. The sum of the weighted inputs and the constant are transformed

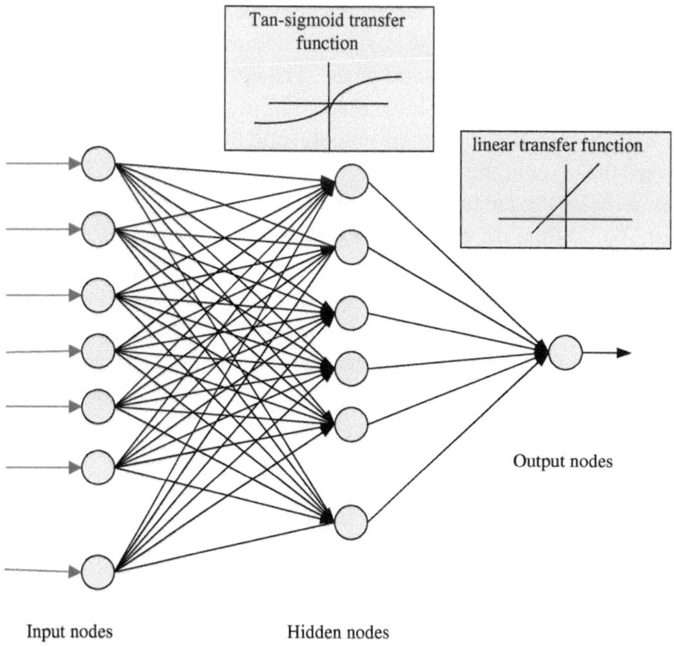

FIG. 3.1 The architecture of the proposed ANN model.

by a function $f(x)$ of the hidden layer. We use the differentiable tan-sigmoid transfer function $f(x)=\frac{1+e^{-x}}{1-e^{-x}}$ to generate their output to the hidden layer. The output of the input layer to hidden node i is

$$o_i=f\left(w_i{}^*X^T+\theta_i\right)=f\left(\sum_{j=1}^{N}\omega_{ij}x_j+\theta_i\right) \tag{3.1}$$

where o_i is the output of the tan-sigmoid transfer function, $w_i=[\omega_{i1},\omega_{i2},\ldots,\omega_{iN}]^T$ is the vector of weights, θ_i is a constant for node i, and $X=[x_1,x_2,\ldots,x_N]^T$ is the vector of input activations.

Similarly, a linear function is employed for the output layer as follows:

$$l(x)=kx+b \tag{3.2}$$

where k is the scope and b is the intercept of the linear function $l(x)$. The output of the only node on the output layer is

$$y=l\left(\sigma^*O^T+\eta\right)=l\left(\sum_{r=1}^{R}\delta_r o_r+\eta\right) \tag{3.3}$$

where $\sigma=[\delta_1,\delta_2,\ldots,\delta_R]^T$ is the vector weights, $O=[o_1,o_2,\ldots,o_R]^T$ is the vector of outputs from the hidden layer, η is the constant, and y is the output of network. After the neural network structure is set, the most important thing is to prepare the training data set and train the network. The training data set must be selected carefully so that the ANN can learn all the relationships through the training process. The designed BP learning phase consists of a forward phase followed by a backward phase. The optimal objective is to minimize the sum of squared errors at the layer at the output side. To do so, the gradient of the error with respect to the weights is found and the weights are adjusted backward toward the layer at the input side. For example, if a neuron is on the hidden layer, it is necessary to calculate the responsibility of the neuron's weights to the final error. To do this, the error at the output neurons is taken and propagated backwards through the current weights, e.g., the same weights used to propagate the activation forward. Adjustment to the neuron's weighting factors is needed if the responsibility exceeds a certain threshold. The training process is then repeated until the specified stopping criteria are satisfied; that is, when the rate of change of the mean squared error is sufficiently small or the mean squared error is sufficiently small, the training stops, and the validation and testing process will be conducted successively [24]. The flow chart of the calculation procedure is shown in Fig. 3.2. The main steps of this procedure are as follows:

(1) Initialize the weights to small random values.
(2) Specify the training vector pair (input—time series of volumes, occupancies, and time stamp and the corresponding output—detected bin volume) from the training set and present the input vector to the inputs of the network.

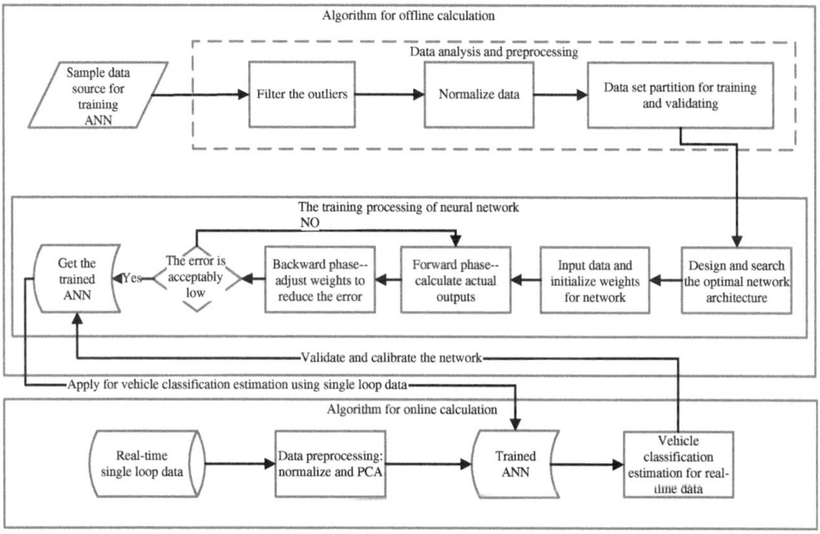

FIG. 3.2 Flow chart of the ANN algorithm.

(3) Calculate the actual outputs (estimated bin volume) as the forward phase.
(4) According to the difference between actual and desired outputs (error), adjust the weights to reduce the difference (in a way that minimizes the error). This is the backward phase.
(5) Repeat steps 2–4 for all training vector.
(6) Repeat from step 2 until the error falls within the threshold value (the error for the entire set is acceptably low).
(7) Stop the procedure of training and apply the network for bin-volume estimation.

After the neural network is properly configured and trained, it is ready for bin volume estimates.

3.3.2 Computer Vision Algorithms to Measure Vehicle Classification Volumes

3.3.2.1 Video-Based Vehicle Detection System Design

The developed VVDC system is designed with six modules and its flow chart is illustrated in Fig. 3.3. The live image stream is carried into the system and digitized to form each frame. The background extraction module is then automatically executed to provide a high-quality background for further application. After the simple system configuration, the vehicle detection and classification algorithms are performed. The information is reciprocally transmitted among these modules shown in Fig. 3.3. Fig. 3.4 provides an illustration of the program user interface.

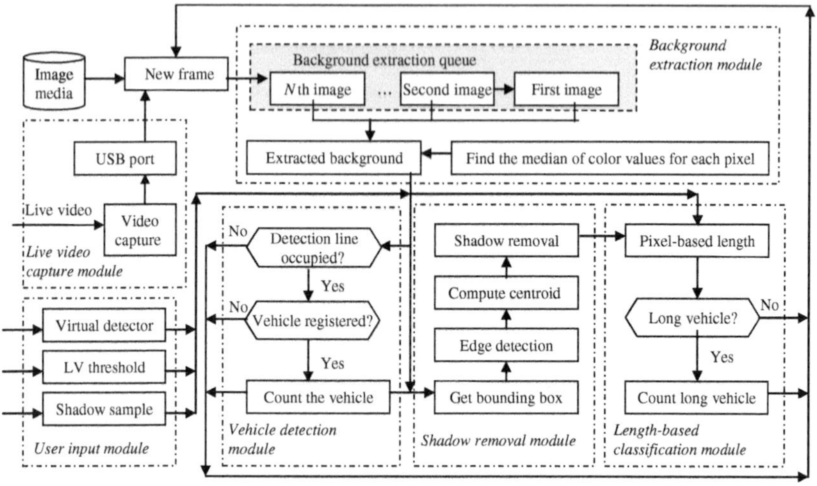

FIG. 3.3 Flow chart of the video-based vehicle detection and classification system.

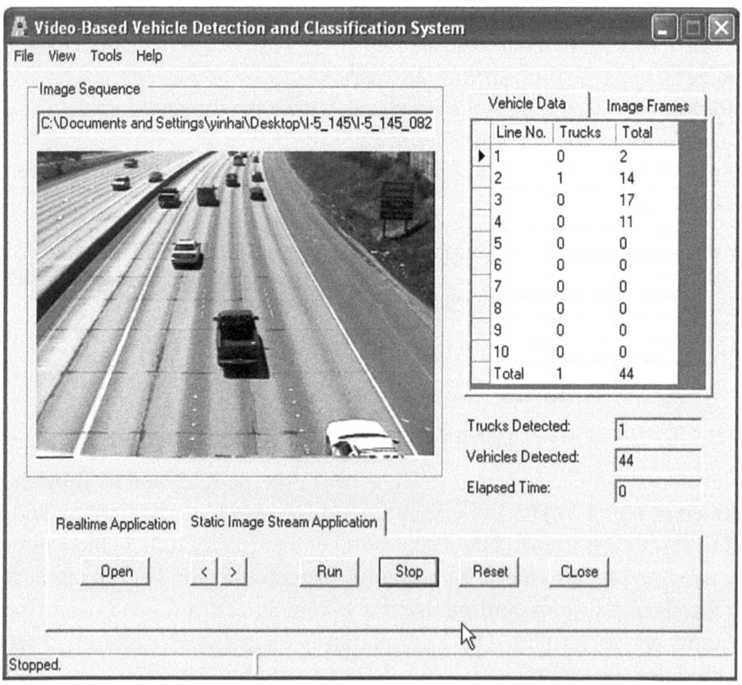

FIG. 3.4 The system user interface.

3.3.2.2 Image Digitization and Background Extraction

In order to strengthen practical applications, a live video capture module is developed to digitize video images in real time from the image stream source, such as video signals from a surveillance cameras or a video cassette player. In this research, a WinTV USB card produced by Hauppauge Digital, Inc. was used to connect a video source to a personal computer. The Microsoft DirectX technology [32] is used in this video capture module. Flexibility in variable transmission rates, image formats, and changeable color configurations of this device extends the system's applicability. In this system, the image format of the Joint Photographic Experts Group (JPEG) and the video frame rate of 20 fps are adopted. When the VVDC system is executed offline, it reads digitized video images from a storage media directly. Based on digitized images, background extraction is conducted to generate a good-quality background for future use. A background image is required to represent the base state of the area under observation for further detection purposes. In terms of traffic detection, it is rarely possible to obtain an image of the observation area that does not contain any vehicles or other foreground objects. Thus, it is necessary to extract the background image from the video stream itself. In this program, the background image is obtained by constructing an image of the median value of each pixel from a collection of images. A color image uses three color channels to represent a pixel's color. These channels in the RGB color space are the Red channel (R), the Green channel (G), and the Blue channel (B). Each channel has a value from 0 to 255 that represents the amount of that color. The color values of a pixel at location (x, y) at the time series t can be expressed as $I_{x,y} = (R_t, G_t, B_t)$. In our study, the median background extraction algorithm is applied, and the median value of each color channel needs to be found for each color pixel. The color values of the pixel at (x, y) in the extracted background $BG_{x,y} = (R_{bg}, G_{bg}, B_{bg})$ can be calculated as follows:

$$BG_{x,y} = \begin{cases} R_{bg} = Median\{R_1, R_2, R_3, \ldots, R_t\} \\ G_{bg} = Median\{G_1, G_2, G_3, \ldots, G_t\} \\ B_{bg} = Median\{B_1, B_2, B_3, \ldots, B_t\} \end{cases} \tag{3.4}$$

where $R_1, R_2, R_3, \ldots, R_t$ is the red channel value of the pixel (x, y) in the image sequence, similar to G and B. By using the median value, it is assumed that the background is predominant in the image sequence. This assumption works reasonably well for freeway applications under moderately congested situations. Fig. 3.5 shows a snapshot of a video scene and the background image extracted. For data collections in locations with higher volumes (which would tend to obscure the background to a greater degree), a background extraction based on the mode of each pixel would be preferable [33]. To adapt dynamically to the luminance change, the background will be extracted and updated periodically. The update cycle may be specified arbitrarily in accordance with weather and lighting conditions.

FIG. 3.5 An example video scene and its background. (A) A snapshot of a video scene; (B) extracted background.

3.3.2.3 *Vehicle Detection*

Before executing vehicle detection and classification algorithms, the basic system configuration needs to be set up. Virtual loop detectors were applied to establish the detection zone and sensors. The concept of a virtual loop is analogous to an inductive loop in that it is placed where vehicles are to be detected. Their different forms were proposed by researchers depending on the specific tasks. In our studies, a virtual loop is comprised of three parts: a registration line, a detection line, and a longitudinal line. This form design not only caters to detection requirements, but also maintains flexibility and simplification in the sensor configuration. Although the virtual detector can be configured in any direction to adapt to detection demands, it should be placed at locations where vehicles are clearly visible with minimal occlusion problems. Each virtual detector will handle the traffic measures on one lane to ensure accurate traffic count and classification data collected. Additionally, the configuration process involves selecting the Automatic Gain Control (AGC) area (light filter box) and sample shadow zone, and the detailed explanation is as follows. Fig. 3.6 demonstrates the system configuration and illustrates a virtual loop discussed above.

One potential disadvantage of using the background subtraction technique is that it does not account for transient lighting changes in the scene [34]. Such effects are often caused by the entrance of a highly reflective vehicle into the scene, such as a large white truck. Before vehicles can be detected, these environmental illumination effects must be accounted for. Correction is performed via the use of an AGC in this study. The AGC is a rectangular area that is placed in a part of the scene where the background is always visible. The average intensity change over this area from the background image can be determined and applied to the entire image to improve accuracy and avoid false vehicle detections:

$$\overline{\Delta int} = \frac{\sum_{A_{agf}} \left(bgint_{i,j} - imint_{i,j}\right)}{A_{agc}} \tag{3.5}$$

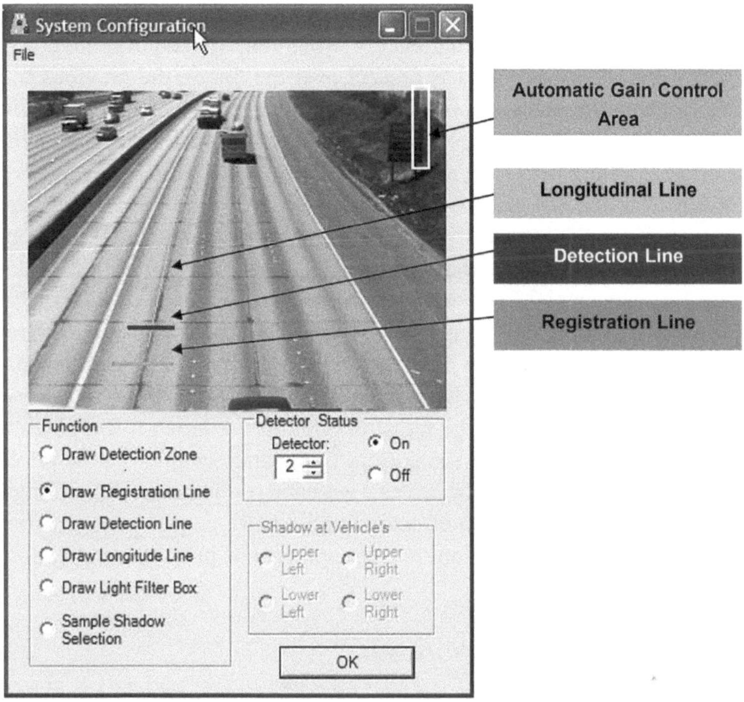

FIG. 3.6 System configuration and components of the virtual detector.

where $\overline{\Delta int}$ is the average intensity difference over the AGC area, A_{agc} is the area of the AGC in number of pixels, $bgint_{i,j}$ represents a pixel intensity in the background image on the interval [0, 1], and $imint_{i,j}$ represents a pixel intensity in the foreground image on the interval [0, 1]. Vehicle detection is then performed using virtual detectors configured by the user. Our vehicle detection algorithm first inspects for vehicles on the registration line:

$$\{p_{i,j} : p_{i,j} \in line\} : d_{i,j} = bgint_{i,j} - imint_{i,j} - \overline{\Delta int} \tag{3.6}$$

where $p_{i,j}$ represents a pixel location, *line* represents the set of all pixels on the registration line, and $d_{i,j}$ is the differenced pixel intensity. We can then define a set C that contains all differenced absolute pixel intensities greater than some threshold t (in this study, a difference of 0.05 was used):

$$C = \{d_{i,j} : |d_{i,j}| > t\} \tag{3.7}$$

If more than 30% of the members of set *line* are also contained in set C, we consider the line to be occupied by a vehicle. To present this fact graphically to the user, the color of the registration line is changed from green to magenta as a visual cue after each detected vehicle. Once detected, the vehicle is processed and classified in one of two ways: entrance detection or exit detection.

In entrance detection, the vehicle is already over the detector when it crosses the registration line and processing occurs when the registration line is first occupied—that is, no vehicle was present over the line in the previous frame. In exit detection, the vehicle is fully in the detector as it is leaving the registration line, and the vehicle is processed upon exiting the registration line—that is, a vehicle was present over the line in the previous frame. When a vehicle is processed, the detection line is inspected for differences in a similar manner that the registration line used previously. Such an entrance-exit detection mechanism double validates detection process and effectively counteracts false alarms resulted from stochastic disturbance, such as slight camera vibrations. Due to regular camera shaking, the image scene shows cyclical fluctuation. The registration and detection lines may be triggered falsely by pixel position changes although they are not occupied by vehicles. In the exit detection process, when the detection line is occupied the registration line is required to keep two continuous states: nonoccupied in the current frame and occupied in the previous frame. A stable current frame becomes a necessary condition for enabling exit detection. This logical decision effectively eliminates negative impacts resulting from regular camera vibrations and strengthens the practical applications of the system.

3.3.2.4 Vehicle Classification

In the application, pixel-based vehicle lengths for each vehicle are obtained. This vehicle length is simply the length along the longitudinal line that is occupied by the vehicle region *V*:

$$len = \sqrt{(e_x - s_x)^2 + (e_y - s_y)^2} \tag{3.8}$$

where s_x, s_y are the start coordinates of the line, e_x, e_y are the end coordinates of the line, and *len* is the pixel-based length of the vehicle. The intersecting points of the longitudinal line with the front and rear edges of the vehicle are needed. To find the intersecting points, a 5×3 mask is used to search along the longitudinal line. If nine of the 15 pixels in the mask are nonbackground pixels, then the center point of the mask is considered to be on the vehicle body. The search starts from the crossing point of the longitudinal line and the detection line and ends when both the front and rear edges of the vehicle are found. The pixel-based length of each vehicle is then compared with a threshold value to determine if it belongs to the SV category or the LV category. Since a vehicle looks different in cameras with different lens and posture settings, the threshold value cannot be a universal predetermined value. The threshold value for each lane is specified by users using the interactive interface with the VVDC system. The length of the longitudinal line of each virtual loop serves as the threshold. Vehicles longer than the longitudinal line are assigned to the LV category. Specifying the length threshold this way provides users with the flexibility for collecting classified vehicle volumes of desired lengths. Note that this detection and classification algorithm is robust to vehicle occlusion in the horizontal direction. Because each virtual loop handles traffic measures on one lane, only the pixel-represented lengths of vehicles along the longitudinal direction will be measured. If vehicles are occluded horizontally, this will not trigger any false detection and classification. Fig. 3.7 shows a snapshot of the system when a vehicle is detected and classified. A red line indicating the detected vehicle length is drawn together with the bounding box describing the rough contour of the detected vehicle.

The benefits of flexibility in algorithm design and ability to use uncalibrated video cameras further expand the usefulness of this prototype VVDC system. The robust algorithms are integrated to address the problems resulting from slight camera vibrations as well as vehicle occlusions in the horizontal direction. Additionally, a simple but effective algorithm addressing shadow identification and removal was developed to improve the system performance. Before presenting the details of the detection and classification algorithms in the methodology section, related studies are introduced briefly. Experimental results and discussion on the performance of this VVDC system are described in the following section. The final section concludes this research effort and proposes further research topics.

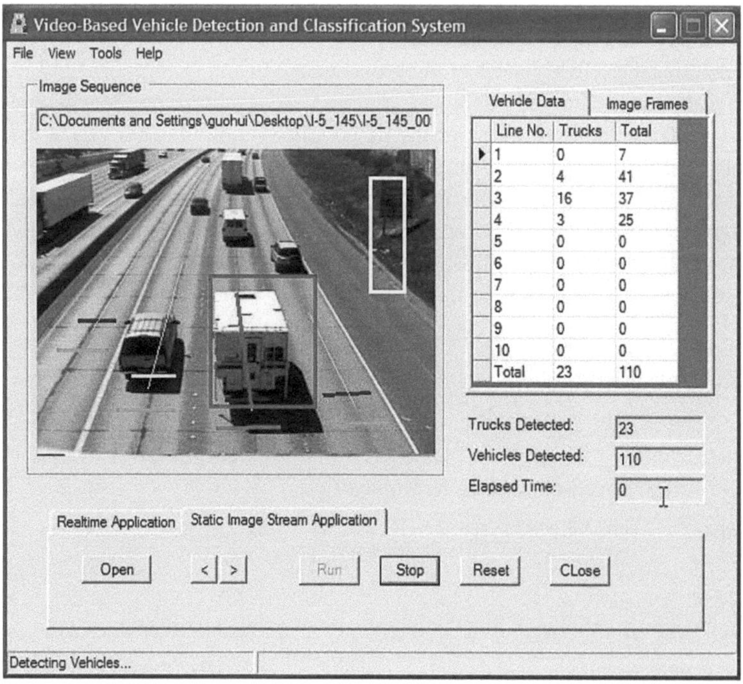

FIG. 3.7 A snapshot of the VVDC system when a vehicle is detected and classified.

3.4 EXPERIMENTAL TESTS AND DISCUSSIONS

3.4.1 ANN Approach Performance Evaluation

To demonstrate the effectiveness of the proposed ANN method, experimental tests were conducted. Two loop stations on I-5 were selected for the test. Details of the stations are summarized in Table 3.2. Each station contains a dual-loop detector for each lane. One of the two single-loops of a dual-loop detector is used as the single-loop data source and the dual-loop measured bin volumes are employed to verify the results from the proposed ANN. The training data set was merely from ES-163R, a dual-loop station at Southbound I-5 under the NE 130th St over bridge. Fifteen days' data, from April 30 to May 13, 1999, were used for the training process. The training data set included single-loop measured volume and occupancy as well as the bin volumes produced by the corresponding dual-loop detectors. As mentioned earlier, the proposed ANN takes all nine 20-s interval measurements over a 3-min period as inputs and outputs the bin volume for the 3-min period. The internal associations among the input data and between the input data and the output data are mined and remembered by the ANN through the training process. To improve the rate of convergence and precision of approximation, the training set was normalized by setting the average of the set to zero and unifying its standard deviation. The Levenberg-Marquardt algorithm was employed by the training process. The entire ANN method was implemented by NeuroIntelligence, a special ANN software tool.

To verify the effectiveness of the proposed ANN and its temporal and spatial transferability, the trained ANN was applied to several test data sets. These test data were collected under various traffic conditions and time periods at the different locations. Interval volumes and occupancies measured by single-loop detectors were used for estimating classified vehicle volumes. The actual bin volumes measured by the corresponding dual-loop detectors were employed to check the results. Figs. 3.8 and 3.9 show the estimated bin volumes and dual-loop observed bin volumes for May 13, 1999 (Thursday) at station

TABLE 3.2 Selected Loop Detectors for Experimental Tests

Station Code	Location	Lane No. (From Right)	Dual-Loop Code[a]	Single-Loop Code[a]
ES-209D	SB I-5 & 156th St. SW	2	_MN__T2	_MN___2
ES-163R	SB I-5 & NE 130th St.	3	_MS__T3	MMS___3

[a] *The WSDOT uses exactly seven characters as loop code to indicate its location and purpose.*

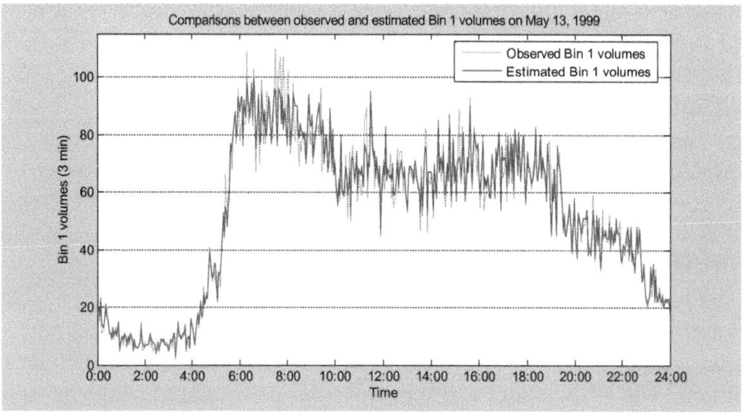

FIG. 3.8 Comparisons between observed and estimated Bin 1 volumes at 3-min level for detector of ES-163R: _MN___2 on May 13, 1999.

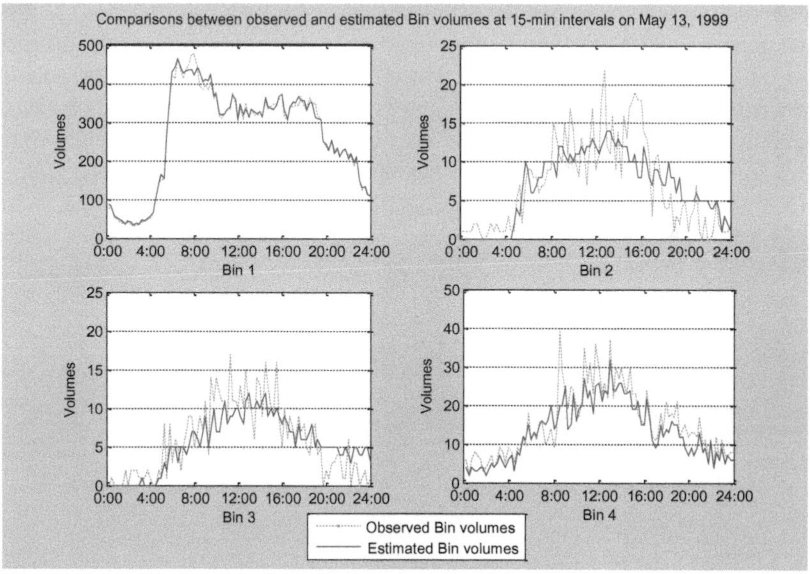

FIG. 3.9 Comparisons between observed and estimated bin volumes at 15-min level for detector of ES-163R: _MN___2 on May 13, 1999.

ES-163R. Due to the low volumes (typically smaller than two per 3-min period) in Bin 2 and Bin 3, results were integrated to 15-min periods for comparisons and the results are shown in Fig. 3.9. In addition to the general comparisons displayed in Fig. 3.9, Fig. 3.8 provides comparisons for the observed and estimated Bin-1 volumes at a higher level of resolution (3-min level). We can see that the two curves overlap and synchronize very well with each other. Since single-loop provides vehicle counts for all four bins, if Bin 1 volume is estimated accurately, the total large-vehicle volume, which contains Bin 2, Bin 3, and Bin 4, can be accurately determined consequently.

The same comparison curves for station ES-209D on May 10, 2004 are shown in Fig. 3.10. We noticed that the estimated volumes for Bin 2 and Bin 3 were significantly larger than the observed Bin 2 and Bin 3 volumes during the afternoon peak period (16:00–18:00). This was probably because of the heavy congestion in this period. When traffic is heavily congested, vehicle speed is significantly lower, which causes unusually long on-times of shorter vehicles mistakenly identified as longer vehicles by the ANN model. Due to the relatively low volumes in Bin 2 and Bin 3, the impacts of such misclassifications were noticeable. Under such seriously congested conditions, a feasible solution is to retrain the ANN model using data collected under the congestion conditions. The ANN model will achieve better performance when the training data sets are collected from situations closer to the application scenario.

To facilitate the comparison between the estimated bin volumes and the observed bin volumes, we define a statistical variable, estimation error as the observed bin volume minus the estimated value for each 3-min period. Means and standard deviations of estimation error in each test case are calculated in order to examine the temporal and spatial transferability of the ANN model. In order to evaluate the strength of association and synchronization between the observed data series and the estimated data series, correlation coefficients (R-value) were also computed. All these results are summarized in Tables 3.3 and 3.4.

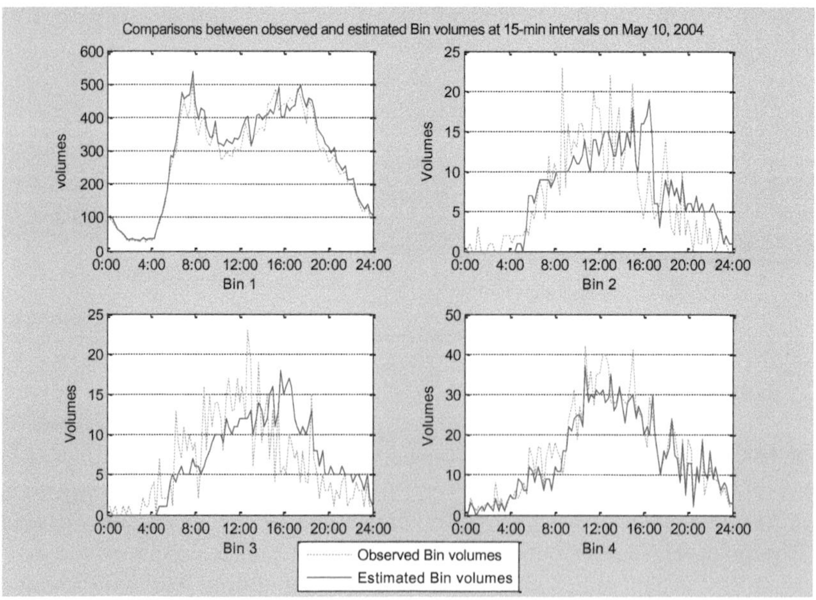

FIG. 3.10 Comparisons between observed and estimated bin volumes at 15-min level for detector of ES-209D: _MN___2 on May 10, 2004.

TABLE 3.3 Statistical Comparisons of Estimation Errors and Correlation Coefficients Between Measured and Estimated Bin Volumes at the Interval of 3 min for Different Days at Station ES-163R

Time Periods	ES-163D											
	Bin 1			*Bin 2*			*Bin 3*			*Bin 4*		
	Mean	*STD*[a]	R[b]	*Mean*	*STD*	R	*Mean*	*STD*	R	*Mean*	*STD*	R
May 13, 1999	−0.47	4.68	0.99	0.07	1.4	0.54	0.09	1.13	0.57	0.44	1.31	0.83
September 6, 1999	−0.33	3.6	0.99	0.09	0.97	0.33	0.19	0.87	0.41	0.06	0.83	0.73
September 7, 1999	−0.29	4.95	0.99	0.38	1.37	0.53	0.6	1.31	0.45	0.53	1.24	0.85
September 8, 1999	−0.34	5.95	0.97	0.22	1.51	0.43	0.46	1.42	0.3	0.44	1.68	0.71
September 9, 1999	−0.59	4.43	0.98	0.24	1.26	0.52	0.49	1.34	0.36	0.33	1.44	0.79

[a]*Standard deviation.*
[b]*R is the correlation coefficient that describes the strength of the association and synchronization between measured and estimated bin volumes.*

TABLE 3.4 Statistical Comparisons of Estimation Errors and Correlation Coefficients Between Measured and Estimated Bin Volumes at the Interval of 3 min for Different Days at Station ES-209D

Time Periods	ES-209D											
	Bin 1			*Bin 2*			*Bin 3*			*Bin 4*		
	Mean	*STD*[a]	R[b]	*Mean*	*STD*	R	*Mean*	*STD*	R	*Mean*	*STD*	R
May 10, 2004	−3.17	8.88	0.96	−0.10	1.49	0.52	−0.03	1.48	0.49	0.29	1.67	0.81
May 11, 2004	−4.04	9.95	0.95	0.02	1.50	0.50	0.04	1.45	0.45	0.47	1.77	0.78
May 12, 2004	−3.89	9.38	0.96	0.02	1.46	0.50	−0.04	1.40	0.43	0.72	1.91	0.76
May 13, 2004	−2.83	10.54	0.94	0.06	1.69	0.51	−0.02	1.41	0.50	0.51	1.90	0.75
May 14, 2004	−1.98	12.55	0.92	−0.09	1.73	0.44	0.19	1.65	0.48	0.58	2.21	0.71
May 15, 2004	−1.76	9.44	0.95	−0.35	1.05	0.36	−0.30	0.92	0.34	−0.29	1.34	0.52
May 16, 2004	−2.17	8.30	0.96	−0.27	0.89	0.36	0.01	1.08	0.38	−0.14	0.92	0.67
May 17, 2004	−1.40	11.31	0.94	0.05	1.54	0.51	0.02	1.55	0.42	0.28	2.00	0.72
May 18, 2004	−1.75	13.94	0.91	0.18	1.81	0.47	0.16	1.61	0.51	0.79	2.40	0.74
May 19, 2004	−2.01	10.49	0.98	0.14	1.40	0.59	0.18	1.27	0.56	0.43	1.98	0.80

[a]*Standard deviation.*
[b]*R is the correlation coefficient that describes the strength of the association and synchronization between measured and estimated bin volume.*

As can been seen from the comparison figures and statistics, this ANN model provided reasonably accurate bin-volumes for the test locations and days, especially for Bin 1 and Bin 4. The comparison curves for different days for station ES-163R indicated that the proposed algorithm yielded favorable results for bin volumes. As shown in Table 3.3, the means of estimation errors for Bin 1 were smaller than 0.6 for all the 5 days tested at station ES-163R, and the standard deviations were less than 6 with the estimation periods of 3 min. The R-values for Bin 1 were greater than 0.97, which indicate that the estimated and observed bin volumes are highly correlated. The estimation accuracy for Bin 1 was the best. Results for other bin volumes were also reasonably good. The result for Bin 4 was better than those for Bin 2 and Bin 3. The reason that the estimation results for Bin 2 and Bin 3 contained larger errors was possibly because of the lower volumes in these two categories. The relatively smaller training samples for these two bins may leave many associations uncaptured in the ANN model, and hence result in larger uncertainty in volume estimations of Bin 2 and Bin 3. However, considering the time lag between dual-loop and single-loop reported data and the possible errors with the dual-loop data, the difference between the estimated bin volumes and the observed bin volumes might have been exaggerated. From a practice perspective, the estimated results are acceptable and applicable.

Estimation accuracies for days from September 6 to 9, 1999 were comparable to that for May 13, 1999, a day from the period when the training data set was collected. This indicates that no significant error was observed when the trained ANN was applied to a different time period at the location from where the training data set was generated. One dual-loop detector at station ES-209D was randomly selected to test the spatial transferability of the proposed ANN model. Ten days' data, from May 10 to May 19, 2004, were collected for the test. Test results are summarized in Table 3.4. We can see that the means and standard deviations of estimation error became larger and R-values also decreased slightly. The largest mean of estimation error was −4.04 for Bin 1 at this station, about 9% of the average Bin 1 volume over a 3-min course. Estimation accuracies for other bins are slightly lower than that of Bin 1, with the largest relative estimation error below 24.6%. Considering that the data set used for this test was collected 5 years after the training data set and at a different location, this test result is very favorable. This concludes that the proposed ANN model is robust and can be applied to different stations on I-5 with reasonable accuracy. However, better accuracy can be obtained if the ANN is tuned with recent data to adapt to the traffic pattern changes. The experimental tests indicated their favorable performance under various traffic operation scenarios. This chapter summarizes our continuous efforts in these promising areas, and contributes greatly to data-driven traffic science and application research.

3.4.2 VVDC System Performance Evaluation

To demonstrate the effectiveness of the VVDC system, two tests with archived video images and one online test with live video data were conducted. For the offline tests two locations were chosen: one from Southbound I-5 near the NE 145th Street over a bridge, and the other from Northbound SR-99 near the NE 41st Street over a bridge. The I-5 test video tape was recorded between 11:30 a.m. and 12:30 p.m. on June 11, 1999. The SR-99 test video tape was taken from 4:00 p.m. to 5:00 p.m. on April 22, 1999. Twelve-min video clips were extracted from the video tapes and digitized. Online test data were from the live video feed link from the WSDOT surveillance video system to STAR Lab at the University of Washington. The camera selected for online testing was the camera shooting Southbound I-5 near the NE 92nd Street over a bridge. The test period of 2:00–5:00 p.m. January 3, 2006 was chosen. These three test collections were selected because they represent wide-ranging application environments: ideal weather conditions with the first test site, and serious shadows with the second test site in the offline test sets. For the live video test, more complex weather and light conditions are chosen to verify the system's robustness. Accompanied with light rain, the slight camera vibration and light reflection generated by the wet road surface challenged the system operation. Two of these three test locations are demonstrated in Fig. 3.11.

Table 3.5 tabulates the results of system evaluation for both offline and online tests at these three sites, including manually observed results (ground-truth data), system operation results, and comparisons between the two. For the offline test at the first location, given the camera location and traffic volume at this site, vehicle occlusion was rare. There were not any shadows that tended to stray into other lanes. Thus, this image set provides an ideal test condition. Test results indicate there is an overall detection error of only 1.06%, and trucks were properly identified approximately 94% of the time. These test results showed encouraging performance of the VVDC system. But one should note that although the system test results were equal to the observed results for truck classification on lane 1 in Table 3.5, this fact does not reflect perfect performance of the system. Comparisons to ground-truth data indicated that there were two mistakes produced by the system: one truck was missed (a false

(A) (B)

FIG. 3.11 Test site situations (A) Northbound SR-99 near the NE 41st Street (B) Southbound I-5 near the NE 92nd Street.

TABLE 3.5 Summary of Results for Both Offline and Online Tests

Time Period		Ground-Truth		System Detected		Comparison Error	
12 min		*Trucks*	*Total Vehicles*	*Trucks*	*Total Vehicles*	*Trucks*	*Total Vehicles*
Location: Southbound I-5 near the 145th Street over a bridge	Lane 1	12	244	12	245	2[a,b]	3[c]
						16.67%[d]	0.82%
	Lane 2	37	335	35	335	2	0
						5.41%	0
	Lane 3	4	409	4	412	0	3
						0	0.73%
	Lane 4	5	149	5	154	0	5
						0	3.36%
	Subtotal	58	1136	56	1146	4	11
						6.89%	1.06%
Location: Northbound SR-99 near the NE 41st Street	Lane 1	15	192	15	194	0	2
						0	1.04%
	Lane 2	7	244	6	245	1	1
						14.28%	0.41%
	Lane 3	8	270	7	270	1	2[e]
						12.5%	0.74%
	Subtotal	30	706	28	709	2	5
						6.67%	0.41%

Continued

TABLE 3.5 Summary of Results for Both Offline and Online Tests—cont'd

Time Period 12 min		Ground-Truth		System Detected		Comparison Error	
		Trucks	*Total Vehicles*	*Trucks*	*Total Vehicles*	*Trucks*	*Total Vehicles*
Location: Southbound I-5 near the 92nd Street over a bridge	Lane 1	5	170	5	173	0	3
						0	1.76%
	Lane 2	5	380	6	389	1	9
						20%	2.36%
	Lane 3	36	378	37	387	3[f]	9
						8.33%	2.38%
	Lane 4	13	388	14	397	1	9
						7.69%	2.31%
	Subtotal	59	1316	62	1346	5	30
						8.47%	2.27%

[a]*Absolute error.*
[b]*One was missed and one was over-counted.*
[c]*Two cars missed and one truck over-counted.*
[d]*Relative percentage error.*
[e]*One vehicle missed and one over-counted.*
[f]*One truck missed and two trucks double-counted.*

(A) (B)

FIG. 3.12 Error investigations: (A) a truck occupying two lanes is measured twice; (B) a misclassified truck with a color of the bed similar to the background color.

dismissal) while another was double-counted (a false alarm). Further investigations of the errors conducted indicate that the major reasons for missing trucks is because the colors of trucks are too similar to the background to have their length measured properly. On the other hand, a truck occupying two lanes was counted twice and led to one truck being over-counted. Fig. 3.12 shows two cases to illustrate these problems. Additionally, several vehicles are over-counted. These false alarms are likely caused by the reflection of vehicle headlights from Northbound I-5 traffic. Furthermore, vehicles changing lanes that did not trigger any of virtual loops resulted in undercounts. For the offline test at the second location, the major purpose was to verify the system operation under serious shadow conditions, as shown in Fig. 3.11A. At this location, vehicle shadows projected into adjacent lanes, which could produce spurious vehicle counts without shadow removal algorithms. Additionally, at this location the traffic flow was interrupted periodically due to signal control at the upstream intersection. The periodical heavy traffic flow could also generate unexpected longitudinal occlusions. The overall results were satisfactory considering that the test conditions were challenging. During the testing period, the overall count error was less than 0.41%, and more than 93% of the trucks presented were correctly recognized. Detailed investigations of the errors indicated that the system handles the negative impacts of shadows effectively. The major problems are caused by sunlight reflection on vehicle bodies and the other reasons similar to that of the first test set.

The online test was conducted using live video from the surveillance camera installed at Southbound I-5 near the NE 92nd Street over a bridge. Selection of this site enabled us to examine the robustness and reliability of the VVDC system when applied to live video images generated from a typical surveillance camera under challenging situations. Compared to an ideal test condition, the image quality of this data set was seriously affected by the low-intensity rain and slight camera vibrations. The moving objects were very small relative to the field of view. Additionally, reflections of vehicle lights on wet pavement became another notable source of disturbance. Therefore, this test is more challenging than the two offline tests introduced earlier. The test results shown in Table 3.5 indicate that the overall accuracy for vehicle count was 97.73% and the truck count accuracy was 91.53%. The performance of the VVDC system was slightly lower in this online test than the offline tests. However, considering that the test conditions were more complicated and challenging, the accuracy levels achieved in this online test are satisfactory. In-depth investigations of the errors revealed that in addition to the typical reasons summarized above, false alarms in vehicle detection were caused mainly by wet pavement reflection. False dismissals were largely due to lane-changing vehicles or vehicles driving on the shoulder without triggering the virtual sensors. Two major causes for vehicle classification errors were longitudinal occlusion and inaccurate estimates of pixel-based length. For some combination trucks with two containers connected by a hitch, the vehicle-length calculation algorithm failed to find the front edge of the vehicle and therefore misclassified it as a SV. Trucks with a trailer or bed in a color similar to the image background experienced similar problems.

3.5 CONCLUSIONS

Due to the size and weight carried, trucks, buses, and recreational vehicles have inferior performance compared to passenger cars. Acquisition of reliable vehicle count and classification data is necessary to establish an enriched information platform and improve the quality of transportation management. However, classified vehicle volumes are not directly measured by the ubiquitously deployed single-loop detectors. Estimating classified vehicle volumes from single-loop outputs is of practical significance. In addition, to utilize the existing video equipment more effectively, it is important to use uncalibrated surveillance video cameras as a cost-effective vehicle detection and classification system to collect SV and LV volumes for each lane on roadways.

In this study, two major approaches were developed to use machine learning mechanisms to establish an ANN model to estimate vehicle classification volumes based on single-loop detector outputs as well as deploy a set of computer vision algorithms to detect SV and LV volumes based on surveillance camera signals. Firstly, we proposed an ANN method in this paper. Vehicle classification categories employed by this study were consistent with the four-bin classification system currently used by the WSDOT dual-loop detection system. To achieve the best bin-volume estimates, a specific neural network was designed and configured for each vehicle category. The proposed ANN was trained and tested using data collected from loop detector stations on I-5 in the greater Seattle area. Our test results indicate that the proposed ANN method worked stably and effectively for the studied stations. The estimated bin volumes were reasonably accurate and can be applied to transportation practice. The temporal and spatial transferability tests showed that the proposed ANN is robust and can be applied to estimate bin volumes during different time periods and at different loop stations on I-5 without introducing significant errors. We can conclude that the proposed ANN is spatially transferable before conducting more tests using data from different routes. Secondly, we developed a set of computer vision algorithms to identify SL and LV volumes based on surveillance camera signals. The effort was taken to develop and integrate several robust algorithms to alleviate the negative impacts resulting from serious shadows, slight camera vibrations, and vehicle occlusion in the horizontal direction. Evaluation results from the three test locations are encouraging. The accuracy for vehicle counting was above 97% for all three tests. The total truck count error was lower than 9% for all three tests. The accuracy for vehicle classification was lower than that for vehicle detection, but is still in the acceptable range. The test results indicate the proposed detection system worked stably and effectively in routine traffic conditions.

Regarding future research directions, although the proposed ANN method produced favorable bin volumes, further improvements to its performance are possible through optimizing its network design and training, especially under heavily congested conditions. Additionally, more accuracy tests using data from different types of road and different areas will also help to explain the spatial transferability of the proposed method. In terms of the video detection system, further improvements are possible through researching on the system operation under congested conditions to extend its application and prevent vehicle misclassification from longitudinal occlusions. Additionally, more robust algorithms addressing light reflection should be investigated and explored to enhance the practical applications of the system.

REFERENCES

[1] TRB (Transportation Research Board), Highway Capacity Manual, TRB, National Research Council, Washington, DC, 2010.

[2] AASHTO (American Association of State Highway and Transportation Officials), AASHTO Guide for Design of Pavement Structures, AASHTO, Washington, DC, 1993.

[3] NCSA (National Center for Statistics and Analysis), Traffic Safety Facts 2003, National Highway Traffic Safety Administration, U.S. Department of Transportation, Washington, DC, 2005.

[4] A. Peters, S. von Klot, M. Heier, I. Trentinaglia, A. Hörmann, H.E. Wichmann, H. Löwel, Exposure to traffic and the onset of myocardial infarction, N. Engl. J. Med. 351 (17) (2004) 1721–1730.

[5] J.J. Kim, S. Smorodinsky, M. Lipsett, B.C. Singer, A.T. Hodgson, B. Ostro, Traffic-related air pollution near busy roads: the east bay children's respiratory health study, Am. J. Respir. Crit. Care Med. 170 (2004) 520–526.

[6] EPA (US Environmental Protection Agency), National Air Quality and Emissions Trends Report, 1999, EPA 454/R-01-004, EPA, North Carolina, 2001.

[7] P. Athol, Interdependence of certain operational characteristics within a moving traffic stream, Highw. Res. Rec. 72 (1965) 58–87.

[8] B. Coifman, S. Dhoorjaty, Z. Lee, Estimating median velocity instead of mean velocity at single loop detectors, Transp. Res. C 11 (3) (2003) 211–222.

[9] Y. Wang, N.L. Nihan, Can single-loop detectors do the work of dual-loop detectors? ASCE J. Transp. Eng. 129 (2) (2003) 169–176.

[10] B. Coifman, Improved velocity estimation using single loop detectors, Transp. Res. A 35 (10) (2001) 863–880.

[11] B. Mikhalkin, H. Payne, L. Isaksen, Estimation of speed from presence detectors, Highw. Res. Rec. 388 (1972) 73–83.

[12] D.J. Dailey, A statistical algorithm for estimating speed from single loop volume and occupancy measurements, Transp. Res. B 33 (5) (1999) 313–322.

[13] A. Pushkar, F.L. Hall, J.A. Acha-Daza, Estimation of speeds from single-loop freeway flow and occupancy data using cusp catastrophe theory model, in: Transportation Research Record: Journal of Transportation Research Board, No. 1457, TRB, National Research Council, Washington, DC, 1994, pp. 149–157.

[14] C. Sun, S.G. Ritchie, Individual vehicle speed estimation using single loop inductive waveforms, J. Transp. Eng. 125 (6) (1999) 531–538.

[15] J. Kwon, P.P. Varaiya, A. Skabardonis, Estimation of truck traffic volume from single loop detector with lane-to-lane speed correlation, in: Transportation Research Record: Journal of Transportation Research Board, No. 1856, TRB, National Research Council, Washington, DC, 2003, pp. 106–117.

[16] N. Mittal, M. Golias, M. Boile, L. Spasovic, K. Ozbay, Estimating truck volumes on state highways—a statistical approach, in: The 84th Annual Meeting of the Transportation Research Board, CD-ROM, Transportation Research Board, National Research Council, Washington, DC, 2005. January 9–13.

[17] P.G. Michalopoulos, Vehicle detection video through image processing: the autoscope system, IEEE Trans. Veh. Technol. 40 (1) (1991) 21–29.

[18] A.H.S. Lai, G.S.K. Fung, N.H.C. Yung, in: Vehicle type classification from visual-based dimension estimation, Proceedings of the IEEE Intelligent Transportation Systems Conference, Oakland, CA, 2001, pp. 201–206.

[19] R.P. Avery, Y. Wang, G.S. Rutherford, in: Length-based vehicle classification using images from uncalibrated video cameras, Proceedings of the 7th International IEEE Conference on Intelligent Transportation Systems, 2004, pp. 737–742.

[20] J. Bonneson, M. Abbas, Video Detection for Intersection and Interchange Control. FHWA/TX-03/4285-1, Texas Transportation Institute, College Station, TX, 2002.

[21] P.T. Martin, G. Dharmavaram, A. Stevanovic, Evaluation of UDOT's Video Detection Systems: System's Performance in Various Test Conditions, 2004. Report No: UT-04.14. Salt Lake City, Utah.

[22] A. Rhodes, D.M. Bullock, J. Sturdevant, Z. Clark, D.G. Candey Jr., in: Evaluation of stop bar video detection accuracy at signalized intersections, Proceedings of the 84th Annual Meeting of Transportation Research Board (CD-ROM), Washington, DC, 2005.

[23] O. Hasegawa, T. Kanade, Type classification, color estimation, and specific target detection of moving targets on public streets, Mach. Vis. Appl. 16 (2) (2005) 116–121.

[24] R. Rad, M. Jamzad, Real time classification and tracking of multiple vehicles in highways, Pattern Recogn. Lett. 26 (10) (2005) 1597–1607.

[25] A.J. Graettinger, R.R. Kilim, M.R. Govindu, P.W. Johnson, S.R. Durrans, Federal highway administration vehicle classification from video data and a disaggregation model, J. Transp. Eng. 131 (9) (2005) 689–698.

[26] J.A.K. Suykens, J.P.L. Vandewalle, B.L.R. De Moor, Artificial Neural Networks for Modeling and Control of Non-linear System, Kluwer Academic Publishers, 1996.

[27] S. Haykin, Neural Networks: A Comprehensive Foundation, second ed., McMaster University, Hamilton, ON, Canada, 1999.

[28] B. Abdulhai, S. Ritchie, Enhancing the universality and transferability of freeway incident detection using a Bayesian-based neural network, Transp. Res. C 7 (5) (1999) 261–280.

[29] S. Hooshdar, H. Adeli, Toward intelligent variable message signs in freeway work zones: neural network model, J. Transp. Eng. 130 (1) (2004) 83–93.

[30] R. Cheu, D. Srinivasan, W. Loo, Training neural networks to detect freeway incidents by using particle swarm optimization, in: Transportation Research Record: Journal of Transportation Research Board, No.1867, TRB, National Research Council, Washington, DC, 2004, pp. 11–18.

[31] H. Teng, Y. Qi, D. Martinelli, Developed incident detection algorithm compared with neural network algorithms, in: Transportation Research Record: Journal of Transportation Research Board, No.1836, TRB, National Research Council, Washington, DC, 2003, pp. 83–92.

[32] Microsoft Inc, Microsoft DirectX Website, http://www.microsoft.com/windows/directx/default.aspx, 2002 (Accessed 16 May 2017).

[33] J. Zheng, Y. Wang, N.L. Nihan, M.E. Hallenbeck, in: Extracting roadway background image: a mode-based approach, Proceedings of the 84th Annual Meeting of Transportation Research Board (CD-ROM), Washington, DC, 2005.

[34] R. Cucchiara, C. Grana, M. Piccardi, A. Prati, Detecting moving objects, ghosts, and shadows in video streams, IEEE Trans. Pattern Anal. Mach. Intell. 25 (10) (2003) 1337–1342.

Chapter 4

Data-Driven Approaches for Estimating Travel Time Reliability

Shu Yang and Yao-Jan Wu

Department of Civil and Architectural Engineering and Mechanics, University of Arizona, Tucson, AZ, United States

Chapter Outline

4.1 Introduction

4.1.1 Significance of Travel Time Reliability

Travel time reliability (TTR) is built on the understanding of travel time and depends on the way of travel time collection. Travel time, obviously, is one of the most important freeway and arterial performance measures, and can be easily understood by both traffic engineers and travelers. Many methods

Data-Driven Solutions to Transportation Problems. https://doi.org/10.1016/B978-0-12-817026-7.00004-7

have been developed to estimate travel times [1–3]. However, travel time estimations or predictions might not give an overview picture of traffic statuses. For example, a 20-min trip from point A to point B might lead to different levels of satisfaction, depending on time of day, day of week, weather conditions, or frequency of traffic accidents. In the past decade, TTR has become as important indicator of roadway performance in addition to travel time. Chen et al. [4] stated that, besides conventional roadway performance measures, such as level of service (LOS), and vehicle miles traveled (VMT), TTR additionally serve as a major indicator of service quality for travelers, and is practically ready to be used to quantify travel costs for personal trips. It is generally accepted that travel cost increases as either travel time increases or TTR decreases. Many researchers have paid attention to the value of knowing TTR. For example, Small et al. [5] translated both mean travel time and the standard deviation of travel time into monetary numbers ($2.60–$8.00 per hour and $10–$15 per hour, respectively), indicating that unreliable trips are more expensive than reliable trips. Additional illustrative examples regarding the value of TTR can be found in Carrion and Levinson's study [6] and the report by Sadabadi et al. [7]. Van Lint et al. [8] stated that travelers prefer to choose reliable routes instead of the shortest ones to improve travel experiences. Therefore, TTR is one of the significant factors that have impacts on travelers' traffic mode choice and route choice. Additionally, TTR has been used not only as a traffic operations performance measure, but also by regional transportation planning authorities to assist in planning and operations at a macroscopic level [9]. These authorities have realized the importance of TTR and focused on using it as a primary measure of roadway congestion, instead of conventional measures such as volume/capacity (V/C) ratio. They realized that using TTR might be a more suitable approach to measuring overall changes of a traffic system.

4.1.2 Definition of TTR

The definitions of TTR vary depending on the purposes of applications, even though TTR has been widely used by transportation planning and operations authorities. In this chapter, we follow the definition by the United States. Federal Highway Administration (FHWA), and the FHWA officially defines "travel time reliability" as "the consistency or dependability in travel times, as measured from day-to-day and/or across different times of the day" [10]. The definition of TTR, from the travelers' point of view, can be interpreted as, "How reliable is the anticipated travel time for my planned trips?" or "In nine out of ten trips, could I arrive at my destination within my expected travel time (90% reliable)?" Obviously, this definition is developed based on the concept of on-time performance. Van Lint and van Zuylen [11] stated that "travel time reliability relates to properties of the (day-to-day) travel time distribution as a function of time of day (TOD), day of the week (DOW), and month of year (MOY), as well as external factors such as weather, incidents, and road work." Therefore, the selection of time period and external factors are composed of

TTR in their definition. The statement also implied that both the supply of traffic-related infrastructure and the demand of traffic can affect TTR. Again, the definition of TTR hereafter in the chapter follows the FHWA's definition, and TTR is estimated from both traffic engineers and travelers' perspectives.

4.1.3 Motivation and Research Questions

As stated above, traffic engineers, authorities, and travelers have begun to realize the importance of TTR. "You cannot manage what you do not measure." Estimating TTR is the first step toward improving it, ensuring on-time arrivals and travel expenses deduction. Therefore, estimating TTR is selected as the focus and this chapter is developed to introduce the basic components of TTR estimation, as well as newly published knowledge of estimating TTR.

Three major research questions are listed in this chapter, including:

(1) TTR measures are insensitive to the selection of probability distribution family, given a certain amount of travel time data [12].
(2) how to measure statistically the accuracy of estimated TTR values.
(3) whether the concept of TTR is extendable from specific road routes to a nonroute-specific (NRS) context (i.e., OD based TTR) and what characters are associated with OD based TTR.

In order to answer the first research question, two hypotheses are specifically proposed, including (1) *Travel times can be fitted with multiple probability distributions and independent on distribution family and* (2) *TTR measures are insensitive to the selection of distribution family* [12]. In order to answer the second question, a pure data-driven technique, the bootstrap, is used to measure the errors of estimated TTR. In order to answer the third research question, three subquestions are developed, including: (1) *how many routes do travelers usually take and what are the TTR values associated with these routes?;* (2) *do statistical differences exist between routes and NRS TTR values?; and* (3) *how to deliver OD-based TTR information since route choice preferences is the additional feature to segment-based TTR?* [3].

4.1.4 Chapter Organization

This chapter consists of five sections. The structure is as follows. Section 4.2 will give a comprehensive literature review regarding travel time distribution estimation, probability distribution family selection, travel time data size selection, and current practices on TTR measures. Section 4.3 will investigate the issues of TTR estimation, including the insensitivity of probability distribution selection and accuracy measurement of TTR. Section 4.4 will expand the concept of TTR from route-specific to NRS contexts, and characterize the OD-based TTR. It will conclude with the research efforts and propose future research topics.

4.2 State of the Art and the Practice

Four basic components are usually considered to measure TTR, including travel time estimation/collection, the selection of probability distribution family, the selection of travel time quantity, and TTR index selection. This section will focus on TTR estimation itself and provide comprehensive literature reviews on the last three components.

4.2.1 Probability Distribution Family Selection for Travel Time Distribution

TTR can be estimated from fitted travel time distributions given a certain amount of travel time data. Travel time distributions are believed to be an important input of estimating TTR. Several previous studies empirically or mathematically demonstrated the relationship between the shape of travel time distribution and traffic flow conditions within a fixed time window. Van Lint and van Zuylen [11] and van Lint et al. [8] described travel time distributions under four different traffic flows: free flow condition, congestion onset, congestion, and congestion dissolve. The relationship was also confirmed in the work by Pu [13] and Guo et al. [14].

Most of the past studies (e.g., [13, 15, 16]) fitted probability distributions by using real-life traffic data. Lomax et al. [17] used Gaussian distributions to fit travel time data. However, the symmetric distributions (i.e., Gaussian distributions) are not the popular distribution family. Instead, skewed probability distributions were most commonly used in previous studies. Polus [18] collected arterial travel time data and used the gamma distribution to fit the data. Al-Deek and Emam [15] found the Weibull distribution was the best distribution to fit their data in a transportation network context by comparing other distributions (e.g., exponential distributions).

Among the skewed probability distributions, the lognormal distribution outperformed other skewed distributions in various traffic flow conditions (e.g., [15, 19, 20]). The lognormal distribution was therefore adopted in the past studies to statistically model the travel time distributions [13], or to serve as a foundation to develop a new TTR measure [21]. These studies used a single probability model to represent the travel time distribution. Guo et al. [14] stated that a probability mixture model outperformed single models, especially under congested conditions. K represents the number of mixture components in a probability mixture model. They first proposed a Gaussian mixture model with $K = 2$, and then calibrated the parameters in the model [22]. Later on, Rakha et al. [23] used a skewed distribution (i.e., lognormal distribution) instead of using the normal distribution as the basic component in the mixture model. K also remained as 2 in their study. Rakha et al. concluded that the skewed mixture model performed the best during peak hours.

In addition to these parametric probability models, nonparametric techniques have been also used to model travel time distribution. Yang et al. [24] used kernel density estimation (KDE) to estimate freeway travel time distribution. They claimed that the major advantage of using KDE was the great flexibility in travel time distribution estimation for various traffic conditions. A new measure was proposed on the basis of KDE-based travel time distributions.

4.2.2 Data Size Selection in Estimating TTR

As stated above, most TTR measures depend on probability distributions. To test and fit the distributions, large quantities of travel time data (often months) must be collected. Table 4.1 summarizes the data size used in several past studies and is ordered by year of publication.

Due to the technique issues and limited availability of traffic data in the 1970s, Polus [18] manually collected 211 travel time samples and then used gamma distributions to fit the data and estimate TTR. As traffic sensors (e.g., radar-based detectors, Bluetooth-enabled sensors, and GPS sensors) became increasingly available, data access becomes more easier and cheaper. This made it possible for van Lint and van Zuylen and Higatani et al. [25] to use 12 months of data to build their travel time distributions [11, 25], while Emam and Al-Deek [16] used only 4 weeks of data to fit their selected probability distributions. Kwon et al. [26] used a nonparametric model to fit data from 256 nonvacation days, claiming that "the sample size is large enough." Yazici et al. [28] used nearly 11 months of data to build the travel time distributions. Lei et al. [32] used only 7 days of floating car data to build travel time distributions. Sumalee et al. [30] used a stochastic cell transmission model to estimate both travel time and TTR. A freeway segment with 30 days of data and an expressway with 7 days of data were included in their study. Carrion and Levinson [29] used 6 weeks of GPS data to quantify the socio-economic and monetary value of TTR [29]. Wakabayashi and Matsumoto [27] conducted comparative studies measuring TTR on highways from the perspective of both travelers and traffic engineers. Two months of data were used to help travelers and traffic engineers understand TTR. Susilawati et al. [31] used GPS-based data to build travel time distributions on urban roadways. Two roads were selected for their study sites, and then travel times were collected from GPS trajectories of 180 and 67 runs. Hojati et al. [33] focused on traffic incidents and evaluating their impact on TTR, but they did not present the data that they used for building travel time distributions and measuring TTR.

People may select the data size for estimating TTR depending on either their experiences or the maximum available data quantity. However, little research is available that provides guidelines for determining the necessary data size for estimating TTR measures.

TABLE 4.1 Summary of Data Size Selection

Authors	Year	Freeway/ Expressway/ Arterial	Quantity	Data Source
Polus [18]	1979	Arterial	211 samples	Floating car
van Lint and van Zuylen [11]	2005	Freeway	12 months	Fixed sensor
Emam and Al-Deek [16]	2006	Freeway	4 weeks	Fixed sensor
Higatani et al. [25]	2009	Expressway	12 months	Fixed sensor
Kwon et al. [26]	2011	Expressway	256 nonvacation days	Fixed sensor
Wakabayashi and Matsumoto [27]	2012	Expressway	2 months	Fixed sensor
Yazici et al. [28]	2012	Expressway	nearly 11 months	Fixed sensor
Carrion and Levinson [29]	2013	Freeway	6 weeks	GPS data
Sumalee et al. [30]	2013	Freeway and expressway	30 days and 7 days	Fixed sensor
Susilawati et al. [31]	2013	Arterial	180 and 67 travel time samples	GPS data
Lei et al. [32]	2014	Freeway	7 days	Floating car
Yang et al. [24]	2014	Freeway	12 months	Fixed sensor
Hojati et al. [33]	2016	Freeway	Not available	Fixed sensor
Yang and Wu [12]	2016	Freeway	12 months	Fixed sensor
Yang et al. [34]	2017	Arterial	12 months	GPS data

4.2.3 Freeway TTR Measures

A considerable amount of research has focused on quantifying TTR from different perspectives. Many popular measures are developed and calculated based on mean or percentiles of travel time, such as the 90th, 95th, buffer index (BI), buffer time, planning time index (PTI), misery index, and Florida Reliability Method. The 90th or 95th percentile travel time is considered to be the "simplest method to measure travel time reliability" [10]. The BI is defined as the additional time that "travelers must add to their average travel time when planning trips to ensure on-time arrival" [10]. Van Lint et al. [8] defined the BI as $\frac{TT90-M}{M}$, where $TT90$ is the 90th percentile travel time and M is the mean travel time. The PTI, which is similar to the BI, represents "how much total time a traveler should allow to ensure on-time arrival" [10]. One definition of the PI used in a study by Pu [13] was $\frac{TT95}{free-flowtraveltime}$, where $TT95$ is the 95th percentile travel time. More general, $TT95$ or $TT90$ can be interchangeably used in either the BI or PI [14].

Moreover, many measures are derived from statistics of the travel time distribution, such as standard deviation and coefficient of variation (CV) of travel times. Van Lint and van Zuylen [11] and Van Lint et al. [8] used skewness and width of the estimated travel time distribution to represent TTR, and they visualized these measures by a "travel time reliability map." These measures are defined in mathematical forms and rephrased in published reports [10, 17, 35] and papers [8, 15, 16, 21, 25]. These TTR measures are easy to understand and interpolated because they are estimated either in minutes (the total travel time or extra time needed to reach destinations) or in percent (level of on time travel performance).

In summary, two categories of TTR measures can be identified, including the moment-based measures (e.g., standard deviation, skewness, kurtosis, and CV) and the percentile-based measures (e.g., the 90th or 95th percentile, BI, and PI) [12]. The moment-based measures are dimensionless, while the percentile-based measures are dimensional indicators.

4.3 Estimating Freeway TTR and Its Accuracy

4.3.1 TTR Measures

Besides travel time data collection and processing, TTR measurement primarily consists of two steps: (1) building travel time distributions with probability distributions; and (2) calculating TTR measures based on the built distributions. The following two sections will give details regarding these steps.

4.3.1.1 Probability Mixture Models

The popular probability distributions for estimating travel time mainly include Gaussian, lognormal, and gamma distributions. Table 4.2 shows the mathematical forms of the three distributions, the first moment, and the second moment of the distribution. These distributions are considered as single distributions and popular in the early stage of TTR measurement. However, researchers have realized the importance of using probability mixture models. A travel time dataset could be fitted with mixture probability models, denoted as $p(x|\Sigma) = \sum_{k=1}^{K} w_i p_i(x|\Sigma_i) = w_1 p_1(x|\Sigma_1) + \cdots + w_K p_K(x|\Sigma_K)$, where $p(x|\Sigma)$ is a continuous probability distribution, $w_i$is the weight of the distribution $p_i(x|\Sigma_i)$, K is the number of distributions, and Σ_i is a set of parameters in the

TABLE 4.2 Statistics of Three Distributions

Distribution Family	Gaussian	Lognormal	Gamma
Probability density function (PDF)	$Gaussian(x;\mu,\sigma) = \frac{1}{\sigma\sqrt{2\pi}} e^{-\frac{(x-\mu)^2}{2\sigma^2}}$	$Lognormal(x;\mu,\sigma) = \frac{1}{x\sigma\sqrt{2\pi}} e^{-\frac{(\ln x-\mu)^2}{2\sigma^2}}$	$Gamma(x;b,\theta) = \frac{1}{\Gamma(b)*\theta^b} x^{b-1} e^{\frac{-x}{\theta}}$
First moment ($\mu^{(1)}$)	μ	$e^{\mu+\sigma^2}$	$b\theta$
Second moment ($\mu^{(2)}$)	σ^2	$(e^{\sigma^2}-1)e^{2\mu+\sigma^2}$	$b\theta^2$

ith distribution. Mixture probability models are simplified to single probability distributions if and only if $K = 1$. Eqs. (4.1)–(4.3) show the expressions for the corresponding mixture models.

$$Gaussian(T\,|\mu,\sigma) = \sum_{k=1}^{K} w_i Gaussian_i(t\,|\mu_i,\sigma_i) \tag{4.1}$$

$$Lognormal(T\,|\mu,\sigma) = \sum_{k=1}^{K} w_i Lognormal_i(t\,|\mu_i,\sigma_i) \tag{4.2}$$

$$Gamma(T\,|b,\theta) = \sum_{k=1}^{K} w_i Gamma_i(t\,|b_i,\theta_i) \tag{4.3}$$

where T is the travel time dataset; w_i is the weight of the ith distribution; μ_i and σ_i are the mean and standard deviation of the ith Gaussian or lognormal distribution, respectively; and b_i and θ_i are the shape and scale of the ith gamma distribution, respectively.

Several parameters must be estimated and calibrated for the mixture models, including the weights (w_i) of individual distributions, the parameters for individual distributions in the mixture models (either μ_i, σ_i or b_i, θ_i), and the number of individual distributions (K). The expectation-maximization (EM) algorithm is the most popular approach to estimate the weights and parameter values in individual distributions when K is given. Rogers and Girolami [36] showed details regarding the parameter estimation in the Gaussian mixture model using the EM algorithm. The parameters in the lognormal mixture model and the gamma mixture model can also be estimated using the EM algorithm using the parameter estimation procedure. Unlike arbitrarily selected initial values for the weights and parameters in individual distributions, the K-means [37] method was used to initialize these weights and parameters, so that the optimal weights and parameters could be estimated efficiently [12].

4.3.1.2 Moment-Based TTR Measures

The most popular moment-based TTR measures include: (1) the two basic moments (the second moment $\mu^{(2)}$ (the variance of travel time), and the third moment $\mu^{(3)}$ (the skewness of travel time)); and (2) two standardized moments (the coefficient of variance and the standardized skewness). Eqs. (4.4)–(4.7) show the expressions for the two basic moments using mixture models. The first moment (a.k.a., the mean value) is also included. Eq. (4.4) shows that the mean value of mixture models is the summation of weighted mean values of individual distributions. The mathematical expression for the Jth moment of mixture model represents $E[(T-\mu)^J]$, and the expression can be rewritten as Eq. (4.5) based on a binomial expansion [38]. Eqs. (4.6), (4.7) show the second moment and third moment of mixture models, respectively.

$$\mu^{(1)} = E[T] = \sum_{k=1}^{K} w_k \mu_k^{(1)} \tag{4.4}$$

$$\mu^{(J)} = E\left[(T-\mu)^J\right] = \sum_{k=1}^{K} \sum_{i=0}^{J} \binom{J}{i} \left(\mu_k - \mu^{(1)}\right)^{J-i} w_k E\left[(T_k - \mu_k)^i\right] \tag{4.5}$$

$$\mu^{(2)} = E\left[(T-\mu)^2\right] = \sum_{k=1}^{K} w_k \left[\left(\mu_k^{(1)} - \mu^{(1)}\right)^2 + \mu_k^{(2)}\right] \tag{4.6}$$

$$\mu^{(3)} = E\left[(T-\mu)^3\right] = \sum_{k=1}^{K} w_k \left[\left(\mu_k^{(1)} - \mu^{(1)}\right)^3 + 3\left(\mu_k^{(1)} - \mu^{(1)}\right) + \mu_k^{(3)}\right] \tag{4.7}$$

where: $\mu^{(i)}$ represents the ith moment about the mean value; $\mu_k{}^{(i)}$ represents the ith moment about the mean value for the kth distribution in mixture models, and the expressions for different choice on $p(x \mid \Sigma)$ are listed in Table 4.2; $E[*]$ is the expression for expectation; and w_k is the weight for the kth distribution in mixture models.

Based on the expressions for the three basic moments, the two standardized moments are derived and shown in Eqs. (4.8), (4.9), respectively. Eq. (4.8) is rewritten using Eqs. (4.4), (4.6), while Eq. (4.9) is rewritten using Eqs. (4.6), (4.7).

Coefficient of variance :

$$\frac{\sqrt{\mu^{(2)}}}{\mu^{(1)}} = \frac{\sqrt{\sum_{k=1}^{K} w_k \left[\left(\mu_k^{(1)} - \mu^{(1)}\right)^2 + \mu_k^{(2)}\right]}}{\sum_{k=1}^{K} w_k \mu_k^{(1)}} \tag{4.8}$$

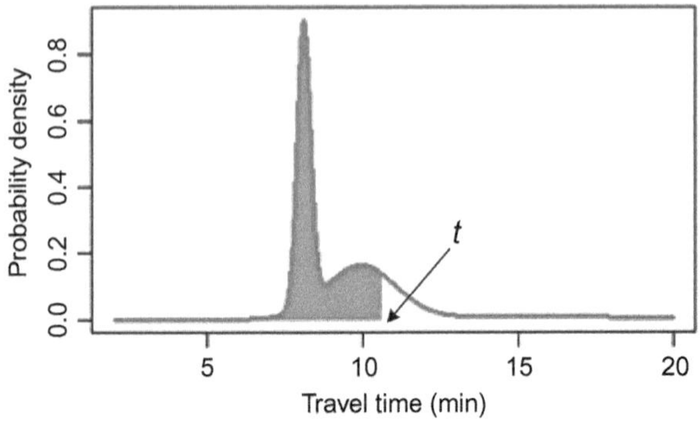

FIG. 4.1 Calculating percentile given a distribution. *(Reproduced with permission from S. Yang, Y.-J. Wu, Mixture models for fitting freeway travel time distributions and measuring travel time reliability, Transp. Res. Record J. Transp. Res. Board (2016) http://doi:10.3141/2594-13, Fig. 1.)*

Standardized skewness :

$$E\left[\left(\frac{T-\mu}{\sigma}\right)^3\right]=\frac{\mu^{(3)}}{\left(\mu^{(2)}\right)^{1.5}}=\frac{\sum_{k=1}^{K} w_k\left[\left(\mu_k^{(1)}-\mu^{(1)}\right)^3+3\left(\mu_k^{(1)}-\mu^{(1)}\right)+\mu_k^{(3)}\right]}{\left(\sum_{k=1}^{K} w_k\left[\left(\mu_k^{(1)}-\mu^{(1)}\right)^2+\mu_k^{(2)}\right]\right)^{1.5}} \tag{4.9}$$

4.3.1.3 *Percentile-Based TTR Measures*

The ith percentile of a probability distribution is located at the value t that represents the area under the probability distribution curve ranging from negative infinity to t where the area equals $i/100$. Fig. 4.1 gives an example of calculating percentiles in an arbitrary distribution. The area of the marked portion is the given i, and t is the corresponding ith percentile. Since mixture models are the summations of weighted individual distributions, the expressions for calculating percentiles of mixture models are technically difficult to derive. The fundamental method of calculating mixture model percentiles is to estimate the area by discretizing the marked area to small portions and summing these portions. This method is also known as a numeric solution for calculating percentile-based TTR.

4.3.2 Insensitivity of Probability Distribution Family Selection

The previous sections gave the details regarding the travel time distribution estimation and TTR measure calculation using mixture probability distributions. However, we believe that "the same data should tell you the same story."

The TTR measures are insensitive to the probability distribution family selection. For instance, given a travel time dataset, the TTR measures calculated from the well-fitted Gaussian mixture models and the lognormal mixture models should be similar. This section examines the insensitivity of probability distribution selection. Two hypotheses were proposed, including (1) *Travel times can be fitted with multiple probability distributions and independent on distribution family and* (2) *TTR measures are insensitive to the selection of distribution family* [12].

Yang and Wu [12] proposed an empirical framework to investigate the issue. Fig. 4.2 depicts the empirical framework. The empirical framework primarily includes the following steps:

- The three mixture models (i.e., Gaussian, Lognormal, and Gamma mixture distributions) are estimated based on a given travel time dataset, and then the moment-based and percentile-based TTR measures are calculated using the estimated probability mixture models.
- The one-sample Kolmogorov-Smirnov (K-S) test is conducted to determine whether the given travel time data statistically follows the probability distributions generated from the mixture models.
- The log-likelihoods, corrected Akaike information criterion (AIC), and Bayesian information criterion (BIC) of the mixture models are also calculated to determine convergence. If the one-sample K-S test suggests that the travel time data statistically follows the three mixture models and the log-likelihoods of those mixture models converge, then the first hypothesis will be accepted; if the TTR measures estimated based on the mixture models remain unchanged, then the second hypothesis will be accepted.
- The number of distributions (K) in the mixture models vary from one to six, meaning that up to six individual distributions in the mixture models are used to fit the estimated travel times for the two cases. The first step is to apply each mixture model to fit the travel times with different K values. Then, the one-sample K-S test is carried out to examine whether the travel times statistically follow the three mixture models.
- The null hypothesis in the K-S test is defined as H_0: *the travel times statistically follow the probability distributions generated from the three mixture models*, while the alternative hypothesis is H_1: *the travel times do not follow the distributions generated from the mixture models*. Two significance levels ($\alpha = 0.01$ and $\alpha = 0.05$) are selected to be the thresholds for indicating whether the travel times follow those generated probability distributions.

The results of the two hypotheses testing are summarized below.

(1) Single probability distributions ($K = 1$) were the inappropriate selection for estimation travel time estimation. Even when $K = 2$, the Gaussian mixture models cannot pass the K-S test, suggesting that a greater K may be appropriate.

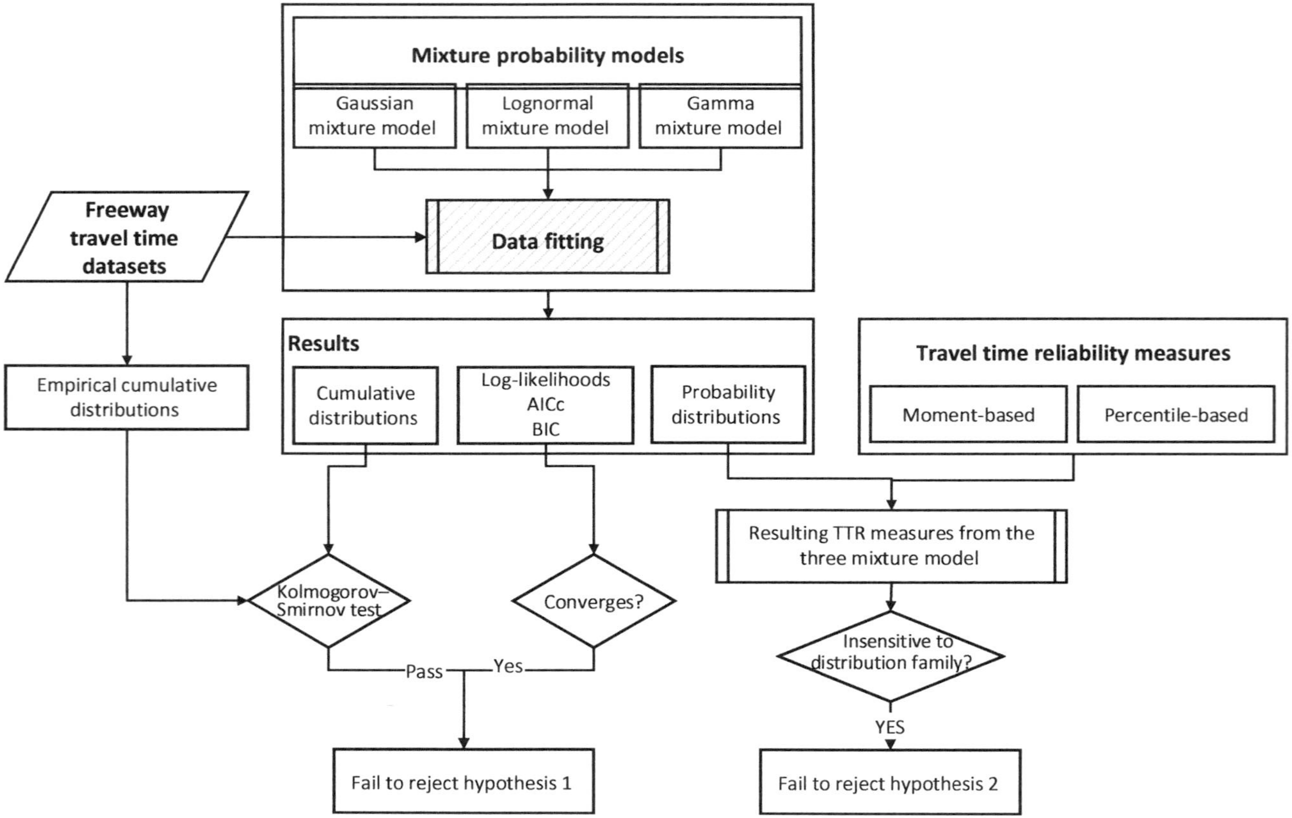

FIG. 4.2 Framework of testing hypotheses. *(Reproduced with permission from S. Yang, Y.-J. Wu, Mixture models for fitting freeway travel time distributions and measuring travel time reliability, Transp. Res. Record J. Transp. Res. Board (2016) http://doi:10.3141/2594-13, Fig. 4.2.)*

(2) Skewed mixture distributions (i.e., lognormal and gamma mixture distribution) outperformed symmetric mixture distributions (i.e., Gaussian mixture distribution), because the symmetric mixture distributions failed to capture the long tail in travel time distributions.

(3) The log-likelihood, AIC, and BIC converged as K increases. Fig. 4.3 clearly shows that the log-likelihoods increased with an increase in K, indicating that the mixture models with a higher value of K can better fit the travel times. Notice that the log-likelihoods converged when K was greater than four, meaning that the addition of further individual distributions may not improve mixture model performance. The convergence curves in Fig. 4.3 can help researchers avoid both underfitting and overfitting, and determine proper selection of K [36].

(4) No significant performance improvement can be found between the three mixture models with higher value of K values.

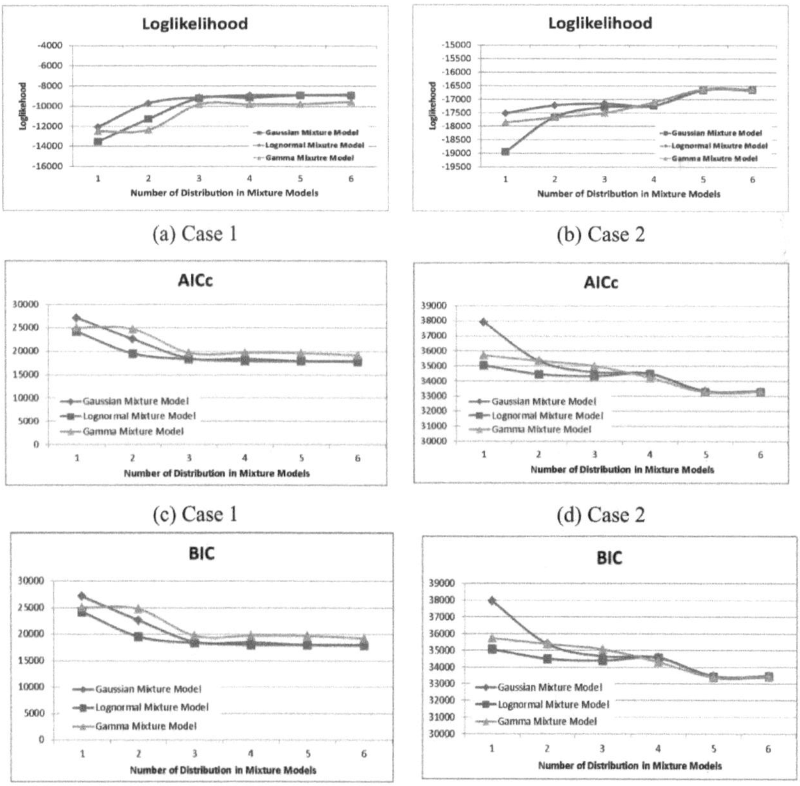

(a) Case 1 (b) Case 2

(c) Case 1 (d) Case 2

(e) Case 1 (f) Case 2

FIG. 4.3 Log-likelihoods of the three mixture models with K lying in [15, 39]. Log-likelihoods (A) Case 1 and (B) Case 2; AIC (C) Case 1 and (D) Case 2; and BIC (E) Case 1 and (F) Case 2.

(5) Given the appropriate level of K, the Gaussian mixture model and the lognormal mixture model performed similarly, and the differences in the log-likelihood values were fairly small (Fig. 4.4).

(6) All of the selected moment-based TTR measures trended closer together with increasing K. No major differences between these moment-based TTR measures could be observed when K was greater than or equal to five. These trending together reflected the log-likelihood convergences, suggesting that if the travel times could be better fitted with the mixture models, the moment-based TTR measures would not change.

(7) Fig. 4.5 shows the 10th, 50th, 90th, and 95th travel time percentiles, and the BI and the PI produced from the three mixture models with variable K. Similarly, the trend in Fig. 4.5 can be observed in the percentile-based TTR measures—that is, no major differences between these percentile-based measures can be observed when K was greater than or equal to five.

4.3.3 Introduction to the Bootstrap

Two popular approaches to estimating the accuracy of an estimator include point estimation and interval estimation. We herein use the standard errors to conduct point estimation, while confidence intervals are used to conduct interval estimations.

Before we jump into the bootstrap, three concepts concerning sampling need to be clarified. Sampling is defined as the statistical procedure of drawing sample data from a population. Resampling is defined as the procedure of drawing sample data from a given observed sample. Resampling can be categorized as resampling with replacement and resampling without replacement. The bootstrap uses the resampling with replacement. For example, given 30 observed travel times on a freeway segment, denoted as $t=(t_1,t_2,t_3,\ldots,t_{30})$, a possibility of resampling with replacement of t could be $t^*=(t_5,t_6,t_{29},t_{21},t_5,t_6,\ldots)$. Note that the number of travel times in t and t^* should be equal.

Mathematically, if a sample $x=(x_1,x_2,x_3,\ldots,x_n)$ is observed, and after B times of resampling with replacement from x, we obtain (x^{*1}, x^{*2}, x^{*3}, ..., x^{*B}), where $x^{*b}=(x_1^{*b},x_2^{*b},x_3^{*b},\ldots,x_n^{*b})$ is called a bootstrap sample and the size of x^{*b} is the same as the original data x. After applying a function of interest $FI(*)$ on the bootstrap samples, it is easy to obtain a series of data $FI(x^{*1})$, $FI(x^{*2})$, $FI(x^{*3})$, ..., $FI(x^{*B})$. The series of data is called bootstrap replication. Then, it is possible to estimate the accuracy of $FI(*)$ using bootstrap replication by evaluating Eqs. (4.10), (4.11).

$$se_{boot}=\sqrt{\frac{\sum_{b=1}^{B}\left(FI(x^{*b})-\overline{FI}\right)^2}{B-1}} \tag{4.10}$$

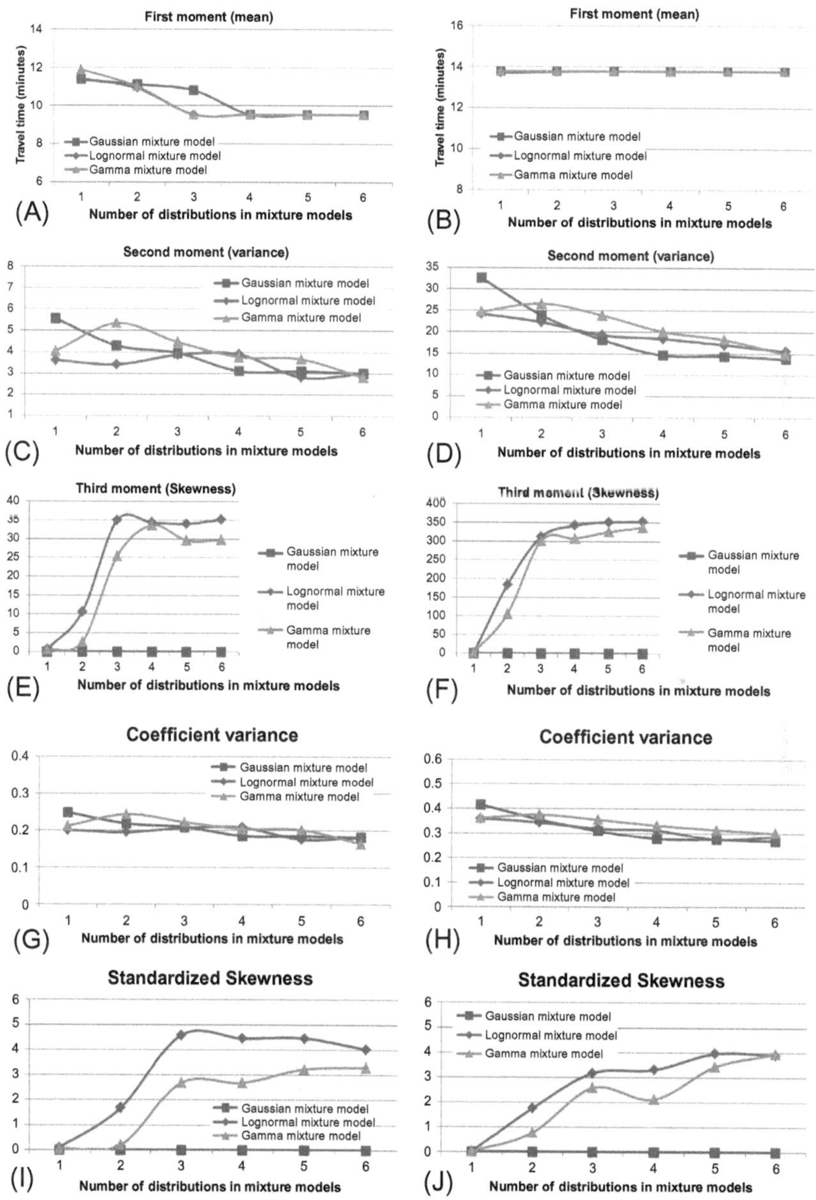

FIG. 4.4 Moment-based travel time reliability measure using the three mixture models: (A) first moment, Case 1; (B) first moment, Case 2; (C) second moment, Case 1; (D) second moment, Case 2; (E) third moment, Case 1; and (F) third moment, Case 2; (G) coefficient of variance, Case 1; (H) coefficient of variance, Case 2; (I) standardized skewness, Case 1; and (J) standardized skewness, Case 2.

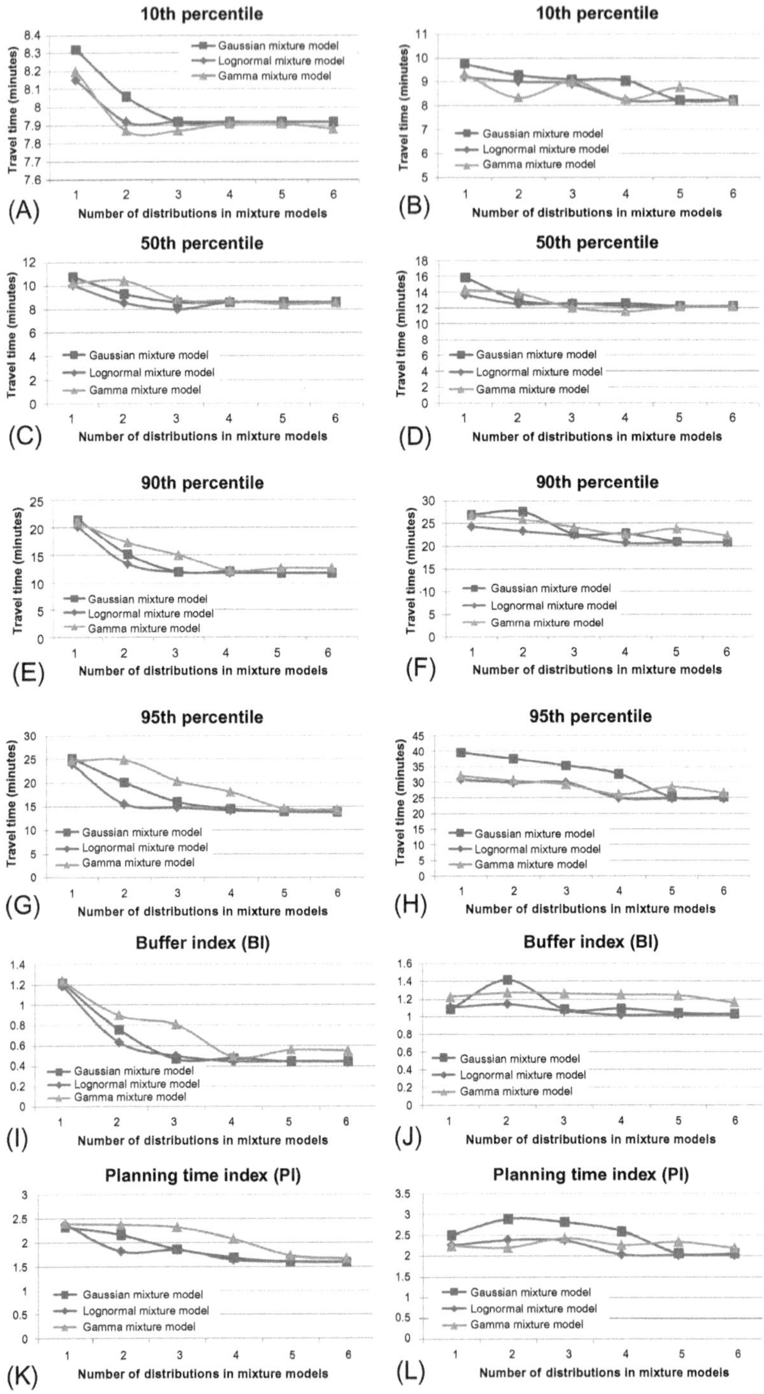

FIG. 4.5 Percentile-based travel time reliability measure using the three mixture models: (A) 10th percentile travel time, Case 1; (B) 10th percentile travel time, Case 2; (C) 50th percentile travel time, Case 1; (D) 50th percentile travel time, Case 2; (E) 90th percentile travel time, Case 1; (F) 90th percentile travel time, Case 2; (G) 95th percentile travel time, Case 1; (H) 95th percentile travel time, Case 2; (I) buffer index, Case 1; (J) buffer index, Case 2; (K) planning time index, Case 1; and (L) planning time index, Case 2.

$$\overline{FI} = \frac{\sum_{b=1}^{B} FI\left(x^{*b}\right)}{B} \tag{4.11}$$

After standard errors of a statistics of interest were estimated, we switch to estimate confidence intervals. Two types of confidence interval are introduced, including percentile intervals and bias-corrected and accelerated (BCa) confidence intervals [39]. Given a raw data $t = (t_1, t_2, t_3, \ldots, t_{30})$ and s selected function of interest $FI(*)$, we are able to obtain the bootstrap replications. Let us denote $FI(x^{*b})$ as $\widehat{\theta(b)}$. The bootstrap replication can be written as $\hat{\theta} = \left(\widehat{\theta(1)}, \widehat{\theta(2)}, \ldots, \widehat{\theta(B)}\right)$. An empirical probability distribution can be generated from $\hat{\theta}$. The two types of percentile intervals are produced by jointly considering the empirical distribution and the significant level (α). Eq. (4.12) is given to calculate the percentile interval of $\hat{\theta}$.

$$\left[\widehat{\theta_{lower}}, \widehat{\theta_{upper}}\right] = \left[\hat{\theta}^{\frac{\alpha}{2}}, \hat{\theta}^{1-\frac{\alpha}{2}}\right] \tag{4.12}$$

where $\widehat{\theta_{lower}}, \widehat{\theta_{upper}}$ are the lower and upper bounds of the intervals, and $\hat{\theta}^{\alpha}$ is the 100 αth percentile of B bootstrap replication.

BCa confidence intervals have been proposed to improve the percentile intervals. The improved percentile intervals can be calculated using Eqs. (4.13)–(4.17)

$$\text{BCa} : \left[\widehat{\theta_{lower}}, \widehat{\theta_{upper}}\right] = \left[\hat{\theta}^{(\alpha_1)}, \hat{\theta}^{(\alpha_2)}\right] \tag{4.13}$$

$$\alpha_1 = \phi\left(\widehat{z_0} + \frac{\widehat{z_0} + z^{(\alpha/2)}}{1 - \hat{a}(\widehat{z_0} + z^{(\alpha/2)})}\right) \tag{4.14}$$

$$\alpha_2 = \phi\left(\widehat{z_0} + \frac{\widehat{z_0} + z^{(1-\alpha/2)}}{1 - \hat{a}(\widehat{z_0} + z^{(1-\alpha/2)})}\right) \tag{4.15}$$

$$\widehat{z_0} = \phi^{-1}\left(\frac{\#\left\{\widehat{\theta(b)} < \overline{\hat{\theta}}\right\}}{B}\right) \tag{4.16}$$

$$\hat{a} = \frac{\sum_{i=1}^{n} \left(\overline{\theta} - \widehat{\theta(b)}\right)^3}{6\left(\sum_{i=1}^{n} \left(\overline{\theta} - \{\widehat{\theta(b)}\right)^2\right)^{1.5}} \tag{4.17}$$

where $\phi(*)$ is the standard normal cumulative distribution function, $\phi^{-1}(*)$ is the inverse function of the standard normal cumulative distribution function, and $z^{(\alpha)}$ is the αth percentile of a standard normal distribution.

Since the bootstrap technique is a compute-intensive technique, a tremendous computing workload is required to evaluate individual bootstrap replications. Parallel computing can be technically used.

4.3.4 Accuracy of TTR Measures

Measuring the TTR accuracy is an application of using the bootstrap technique. The function of interest $FI(*)$ can be any TTR measures—for example, $FI(*)$ could be the PI, and then the PI can be plugged into Eqs. (4.10), (4.11). Fig. 4.6 shows the four steps starting from travel time data collection to TTR accuracy measurement using the bootstrap technique.

For demonstration purposes, we used the traffic data provided by the Missouri Department of Transportation, US The details of three case studies are listed in Table 4.3. The travel time data used for Case study 1 was associated with TOD = 12 p.m. and DOW = Thursday under free-flow traffic. Table 4.4 summarizes the outputs of the TTR accuracy measurement framework using 23 weeks of travel time data. The expected values of the four TTR measures were generally used to report TTR. Standard errors, 95th percentile confidence intervals, and 95th BCa confidence intervals were used to examine the accuracy of these expected values. For example, the expectation of the PI was 1.0377, indicating that travelers should plan on approximately 4% additional time to ensure on-time travel in Case study 1. The standard errors of PI were calculated using 1000 bootstrap replications. The standard error for the PI was relatively low (0.0026), and thus the expected PI value was considered to be statistically accurate. Meanwhile, the 95th percentile interval for the PI lay within the range [1.0327, 1.0431], and the 95th BCa percentile interval was [1.0328, 1.0432]. The two types of percentile intervals were similar and small, suggesting that the expected PI value was considered statistically accurate.

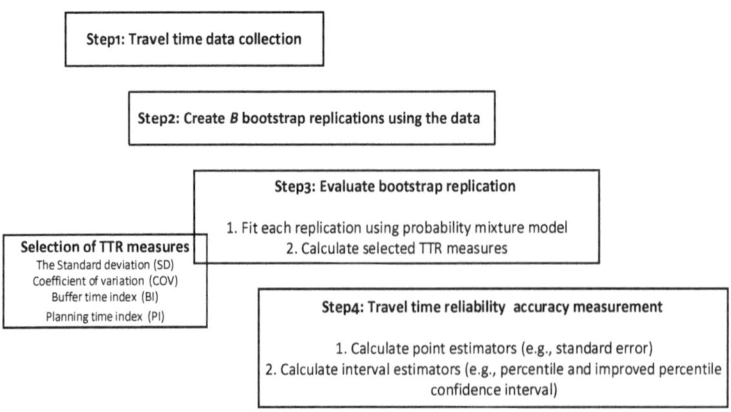

FIG. 4.6 Framework of measuring the accuracy of travel time reliability.

TABLE 4.3 Optimal Quantity Case Studies

Case Study	Location	Length (Miles)	TOD, DOW	Traffic Condition	Average Travel Time (Min)
1	I-64 WB from Route K to Prospect Road	6.5	12 pm, Thursday	Free flow	6.5
2	I-270 SB from D-Page Ave to Dougherty Ferry Road	8.1	4 pm, Thursday	Recurrent congestion	13.1
3	I-270 SB from D-Page Ave to Dougherty Ferry Road	8.1	3 pm, Thursday	Transition	11.9

TABLE 4.4 Case Study 1: 23 Weeks of Data

	TTR Measures			
	SD	*COV*	*BI*	*PI*
Expected value	5.659	0.0146	0.0295	1.0377
Standard error	1.534	0.0039	0.0019	0.0026
95th percentile interval	[2.870, 8.659]	[0.0074, 0.0223]	[0.0258, 0.0333]	[1.0327, 1.0431]
95th BCa	[3.066, 8.749]	[0.0078, 0.0226]	[0.0259, 0.0334]	[1.0328, 1.0432]

Similarly, the proposed steps were applied to Case studies 2 and 3. Tables 4.5 and 4.6 show the results of the accuracy measures. For example, the expected value of the PI (1.6839 in Case study 2) suggests that travelers should plan on an additional nearly 70% travel time to guarantee on-time arrival in Case study 2. The standard error for the PI was greater than the corresponding

TABLE 4.5 Case Study 2: 23 Weeks of Data

	TTR Measures			
	SD	*COV*	*BI*	*PI*
Expected value	165.559	0.1890	0.5060	1.6839
Standard error	62.948	0.0715	0.0294	0.0410
95th percentile interval	[30.897, 262.721]	[0.0354, 0.2962]	[0.4455, 0.5641]	[1.6035, 1.7670]
95th BCa	[39.203, 265.163]	[0.0433, 0.2988]	[0.4454, 0.5629]	[1.6037, 1.7679]

TABLE 4.6 Case Study 3: 23 Weeks of Data

	TTR Measures			
	SD	*COV*	*BI*	*PI*
Expected value	76.160	0.1265	0.3319	1.4738
Standard error	28.529	0.0471	0.026	0.0348
95th percentile interval	[36.564, 126.120]	[0.0616, 0.2083]	[0.2761, 0.3822]	[1.4079, 1.5411]
95th BCa	[44.126, 134.934]	[0.0739, 0.2202]	[0.2729, 0.3803]	[1.4083, 1.5418]

one in Case study 1. The two types of the percentile intervals were relatively large, approximately [1.6036, 1.7679]. A similarly large range was also noticed in Case study 3. The expected value for the PI was 1.4738, with the standard error 0.0348 and the two types of the percentile intervals were approximately [1.4079, 1.5418].

4.3.5 Optimal Quantity of Travel Time

When the accuracy of TTR measures can be estimated given a certain amount of travel time data, traffic professionals and researchers may try to test different amount of travel time data to check if the TTR accuracy meets the needs of applications and research. After repeating the four steps in the previous section with different amount of travel time data (from 4 to 60 weeks of travel time data), optimal travel time data quantities (measured in weeks) were empirically determined in our case studies. We make the following suggestions below for determining the optimal data size of travel time for TTR measurement.

(1) Free-flow condition (Case study 1): even a small portion of travel time data can produce statistically accurate TTR measures. In our study, the optimal quantity for the free-flow condition could be as little as 2 weeks. We recommend 4 weeks of travel time data instead, because TTR measures can be more stable when using this quantity of data.
(2) Congested condition (Case study 2): the TTR measures generally had high statistical accuracy with more data included. However, the increase in data did not guarantee an increase in the accuracy. We recommend 40 weeks of travel time data to be the optimal data size in our study; the accuracy of the TTR measures would not be significantly improved with more than this amount of data.
(3) Transition condition (Case study 3): we recommend 35 weeks of data would be enough to compromise between the measures of accuracy and data quantity.

4.4 From Segment-Based TTR to OD-Based TTR

Intelligent transportation systems (ITS) sensors (e.g., loop and radar-based sensors, Bluetooth-enabled sensors, Wi-Fi-enabled sensors) have been installed along roadways to collect traffic data. These data have been utilized for travel time and TTR estimation. Most of the previous studies can be considered as segment-based, because travel times are estimated on segment s of interest where data is available. Using vehicles equipped with GPS-enabled devices have been increasingly popular approach to collect travel times over the past decades [40]. Fig. 4.6 suggested four steps to measure TTR and its accuracy. Traffic professionals may utilize the data collected from fixed location ITS sensors or GPS devices to estimate travel times first, and then estimate TTR measures [12, 24]. The steps imply that roadway segments have to be selected prior to estimating travel times and TTR measures. These TTR estimations are accordingly named segment-based TTR because travel time collection and TTR estimations are conducted on specified segments. Transportation agencies are responsible for reporting segment-based TTR measures to upper-level agencies and the public by utilizing the data collected from fixed location ITS sensors. It is generally accepted that the GPS-enabled data can be used for estimating either segment-based or OD-based travel times. TTR is defined and built on large amount of historical travel time data. Since travel time can be generally categorized as segment-based and OD-based travel time, TTR can be categorized as segment-based and OD-based TTR [34].

4.4.1 Significance of OD-Based TTR

Few studies, however, have investigated and explored the issues and benefits of OD-based TTR estimations. This section aims to develop and uncover OD-based TTR estimations. Instead of segment specification, pairs of origin and destination must be specified before estimating both OD-based travel times and TTR measures. Therefore, the concept of OD-based TTR is defined as "how reliable is your trip from an origin to a destination?" [34]. Travelers may take various routes to reach their destinations due to current traffic conditions or personal preferences. Therefore, OD-based TTR measures could be estimated in two ways: (1) evaluating TTR measures for individual routes; or (2) evaluating NRS TTR measures. Due to the concept differences between segment-based and OD-based TTR, three issues can be identified when implementing OD-based TTR measures [34], including

(1) How many routes travelers usually take and what are the TTR values associated with these routes?
(2) Do statistical differences exist between routes and NRS TTR values?
(3) How to deliver OD-based TTR information since route choice preference is the additional feature to segment-based TTR?

4.4.2 OD-Based TTR Measurement

For demonstration purposes and investigating the three issues above, 6 months of GPS-based data were collected from taxicabs operated in the City of Kunshan, China. The origin and destination are plotted in Fig. 4.7. The origin was selected as a large living neighborhood, while the Kunshan South Railway Station was selected as the destination. People usually commute from the living neighborhood to the railway station to take trains to their workplaces. The marked dash line in Fig. 4.7 is the shortest path between the origin and destination.

After processing and analyzing the GPS-based data collected from taxicabs, three routes were identified in Fig. 4.8 as travelers' preferred routes. The details of these routes are as follows [34].

- Route 1 was spatially close to the shortest path shown in Fig. 4.7, and the traveling distance on Route 1 was 13.8 km (8.6 miles). The roadways in Route 1 were in an urban environment, and the most of the roadways are equipped with traffic signals.
- Route 2 was a little longer than Route 1. The traveling distance on Route 2 was 14.3 km, or 8.9 miles. Route 2 was alternatively selected by taxicab drivers. One of the obvious reasons was that the majority of roadways in Route 2 were less congested than those in Route 1. Most of the roadways in Route 2 were also in an urban environment with traffic signals.

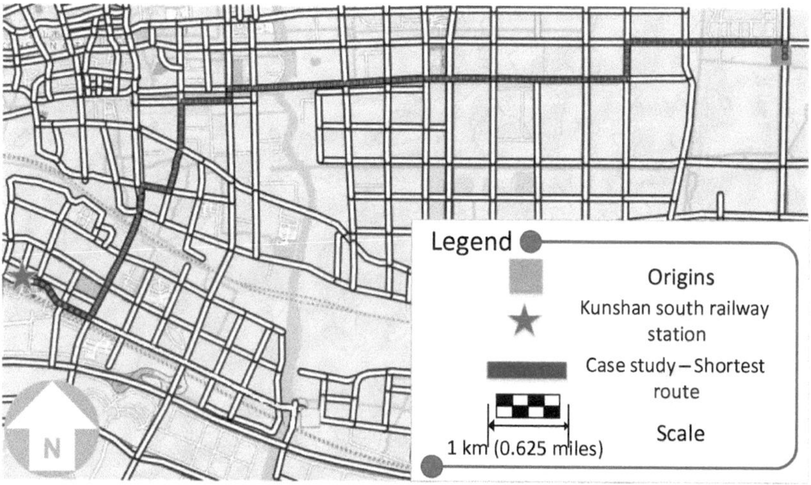

FIG. 4.7 Origin and destination, and its shortest routes. *(Reproduced with permission from S. Yang, C. An, Y.J. Wu, J. Xia, Origin-destination based travel time reliability, Transp. Res. Record J. Transp. Res. Board, 2643 (2017). https://doi.org/10.3141/2643-16, Fig. 4.3.)*

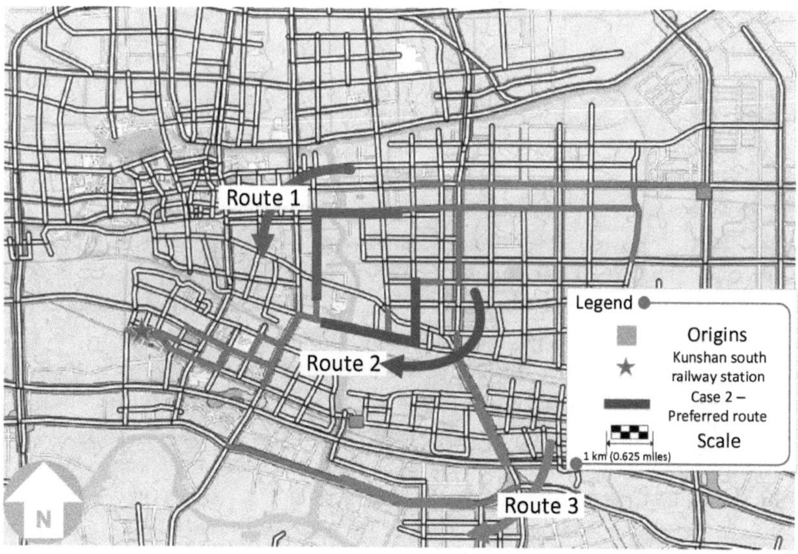

FIG. 4.8 Three preferred routes, case study. *(Reproduced with permission from S. Yang, C. An, Y. J. Wu, J. Xia, Origin-destination based travel time reliability, Transp. Res. Record J. Transp. Res. Board, 2643 (2017). https://doi.org/10.3141/2643-16, Fig. 4.4.)*

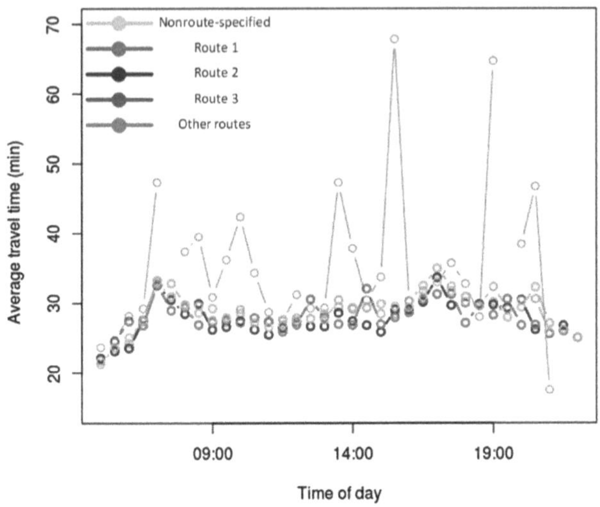

FIG. 4.9 Average travel times by preferred route [34].

- Route 3 was 21.1 km (13.1 miles), which was approximately 50% longer than both Routes 1 and 2. However, the major roadways in Route 3 were expressways without traffic signals.

Any other routes differed from the primary three preferred routes were identified as *Others*. Fig. 4.9 showed that only a few travelers took other routes to the station.

After separating the GPS-based data into three routes, the three groups of travel times were calculated. Two TTR measures (i.e., standard deviation and BI) were selected to calculate. Table 4.7 only lists the data with the time period 6:00–6:30 a.m. The standard error was also calculated to present the accuracy of the two TTR measures. For example, approximately 44.7% taxicabs took Route 1 from the neighborhood to the station. The two TTR measures were 2.21 and 0.13, respectively. The accuracy measurements of the two TTR measures were 0.82 and 0.06, respectively, suggesting that the two TTR measures were statistically accurate. In addition to separating into three preferred routes, we also treated the data as a group, which is NSR in Table 4.7. The two TTR measures and the accuracy associated with the two TTR measures were calculated as well. Computational analysis of variance (ANOVA) was also performed to examine statistically whether the average travel times on the three routes were statically equal. All of the results are listed in Table 4.7. More examples can be found in the work by Yang et al. [40].

Therefore, we can answer the first two research questions regarding the OD-based TTR measures. The answers are summarized below [34].

Overall, no statistically significant differences existed between route-specific and NRS TTR values for most of the time periods. This suggested

TABLE 4.7 TTR Measures and Their Accuracy [34]

			Standard Deviation (SD)			Buffer Index (BI)		
	Time Period	Sample (%)	*SD* (Standard Error)	a	b	*BI* (Standard Error)	a	b
Route 1	6:00–6:30 a.m.	21 (44.7%)	2.21 (0.82)	1.33 4.03	*Y*	0.13 (0.06)	0.03 0.21	*Y*
Route 2		13 (27.7%)	2.69 (1.41)	0.13 4.77		0.13 (0.05)	0.02 0.16	
Route 3		5 (10.6%)	2.66 (1.90)	1.53 7.78		0.11 (0.04)	0.08 0.22	
Others		8 (17.0%)	2.47 (1.89)	−1.03 5.17		0.19 (0.09)	0.08 0.38	
NSR		47 (100%)	2.75 (0.93)	1.01 4.06		0.19 (0.04)	0.14 0.27	

[a] *Means corresponding 95th percentile intervals.*
[b] *Whether the average travel times are statistically equal.*

that few statistically significant differences can be found between route-specific and NRS TTR measures and the NRS TTR measures can serve as alternatives for route-specific TTR measures. However, significant differences can be still found in some cases. For example, during 2:00–2:30 p.m., the NRS TTR measures were statistically different to the route-specific TTR measures.

Smaller route TTR measures suggested that taking this preferred route may have advantages in terms of TTR, since the average travel times were similar (see Fig. 4.9). Travelers may utilize this information to select routes with smaller TTR measure as more reliable routes.

4.4.3 OD-Based TTR Information Delivery

Segment-based TTR measures usually provide numeric numbers only because route information is already implied. However, OD-based TTR measures could provide both numeric numbers and route choice preferences for travelers. The case study showed that: (1) distance-based shortest paths can be preferred routes; (2) average travel times by preferred route slightly differed; and (3) TTR measures by preferred route might be statistically different [34].

Therefore, two ways of OD-based information delivery can be adopted: (1) both numeric TTR values and route preference can be jointly published; and (2) if preferred routes could not be identified, NRS TTR measures can alternatively represent route-specific TTR measures, and distance-based shortest paths could serve as the references of preferred routes.

4.5 Conclusion and Recommendations

Due to the extensive availability of various traffic sensors on roadways, traffic data collection has become easier, cheaper, and largely available. TTR, which benefits from data availability, has attracted increasing attention in recent years, and is often listed as one of the major roadway performance and service quality measures. Measuring TTR is the first step toward improving TTR, ensuring on-time arrivals, and reducing travel costs. Four basic components are usually considered to measure TTR, including travel time estimation/collection, quantity of travel time selection, probability distribution selection, and TTR index selection. This chapter empirically proves the concept that "the same data tells you the same story" and TTR measures are insensitive to the probability distribution selection. Two hypotheses were proposed: including (1) *Travel times can be fitted with multiple probability distributions and independent on distribution family and* (2) *TTR measures are insensitive to the selection of distribution family*. An experimental framework was used to investigate the two hypotheses empirically. The results suggested that: (1) travel time can be precisely fitted using mixture models with higher value of K instead of using single probability distributions, regardless of the probability distribution family in mixture models; and (2) the second hypothesis was tested through comprehensive investigation of moment-based and percentile-based reliability measures, suggesting that the TTR measures were insensitive to the selection of probability distribution family.

This chapter also covers an additional component beyond the basic components, which is the estimation of the accuracy of TTR measures. The bootstrap technique, which is a data-driven technique and is based on resampling with replacement, plays an important role in estimating the accuracy. The accuracy estimations provide more general pictures for estimated TTR rather than a single TTR value. The results of the accuracy measurements of the TTR measures primarily showed that (1) the estimates of the TTR measures under free-flow conditions were more accurate than those under congested and transition conditions, and (2) the estimates of the TTR measures under congested conditions had the least accuracy, suggesting that more data may be used to improve the accuracy.

It is generally accepted that the GPS-enabled data can be used for estimating either segment-based or OD-based travel times. TTR is defined and built on large amount of historical travel time data. Since travel time can be generally categorized as segment-based and OD-based travel time, TTR can be categorized as segment-based and OD-based TTR. Three sub-questions are developed to characterize the OD-based TTR, including: (1) *how many routes travelers usually take and what are the TTR values associated with these routes?;* (2) *do statistical differences exist between routes and NRS TTR values?; and* (3) *how to deliver OD-based TTR information since route choice preferences is the additional feature to segment-based TTR?* The results showed that no statistically significant differences existed between route-specific and NRS TTR measures for most of the time periods. Statistically significant differences can be still found in some time periods (e.g., rush hours). Travelers may take advantage of these differences to choose a more reliable route.

We suggest the following recommendations to estimate TTR statistically accurately.

(1) Lognormal mixture models with $K = 3$ are preferred to estimate travel time distribution. However, the Gaussian mixture models can alternatively be used.
(2) It is suggested that both single TTR values and TTR interval values are reported, as well as the travel time data quantity.
(3) OD-based TTR estimation consists of two parts: route information and TTR values. It is recommended to report both. When route information is unavailable, the shortest paths and NRS TTR can be used.

References

[1] S. Lim, C. Lee, Data fusion algorithm improves travel time predictions, IET Intell. Transp. Syst. 5 (4) (2011) 302.

[2] L. Sun, J. Yang, H. Mahmassani, Travel time estimation based on piecewise truncated quadratic speed trajectory. Transp. Res. A Policy Pract. 42 (1) (2008) 173–186, https://doi.org/10.1016/j.tra.2007.08.004.

[3] S. Yang, Y.-J. Wu, Z. Yin, Y. Feng, Estimating freeway travel times using general motors model. Transp. Res. Record J. Transp. Res. Board (2016), https://doi.org/10.3141/2594-12.

[4] C. Chen, A. Skabardonis, P. Varaiya, Travel-time reliability as a measure of service, Transp. Res. Record J. Transp. Res. Board 1855 (2003) (2003) 74–79.

[5] K.A. Small, R. Noland, X. Chu, D. Lewis, Valuation of Travel-Time Savings and Predictability in Congested Conditions for Highway User-Cost Estimation, Retrieved from, http://ntl.bts.gov/lib/16000/16200/16215/PB2000103014.pdf, 1999.

[6] C. Carrion, D. Levinson, Value of travel time reliability: a review of current evidence. Transp. Res. A Policy Pract. 46 (4) (2012) 720–741, https://doi.org/10.1016/j.tra.2012.01.003.

[7] National Academies of Sciences, Engineering, and Medicine, Value of Travel Time Reliability in Transportation Decision Making: Proof of Concept—Maryland. The National Academies Press, Washington, DC, 2014. https://doi.org/10.17226/22280.

[8] J.W.C. Van Lint, H.J. van Zuylen, H. Tu, Travel time unreliability on freeways: why measures based on variance tell only half the story. Transp. Res. A Policy Pract. 42 (1) (2008) 258–277, https://doi.org/10.1016/j.tra.2007.08.008.

[9] K. Lyman, R.L. Bertini, Using travel time reliability measures to improve regional transportation planning and operations, Transp. Res. Record J. Transp. Res. Board 2046 (2008) (2008) 1–10.

[10] Texas Transportation Institute, & Cambridge Systems, I. (2006). Travel Time Reliability: Making It There On Time, All the Time. Federal Highway Administration Office of Operations. Retrieved from, http://www.ops.fhwa.dot.gov/publications/tt_reliability/

[11] J.W.C. Van Lint, H.J. van Zuylen, Monitoring and predicting freeway travel time reliability: using width and skew of day-to-day travel time distribution. Transp. Res. Record J. Transp. Res. Board 1917 (1) (2005) 54–62, https://doi.org/10.3141/1917-07.

[12] S. Yang, Y.-J. Wu, Mixture models for fitting freeway travel time distributions and measuring travel time reliability. Transp. Res. Record J. Transp. Res. Board (2016). https://doi.org/10.3141/2594-13.

[13] W. Pu, Analytic relationships between travel time reliability measures, Transp. Res. Record J. Transp. Res. Board 2254 (2011) (2012) 122–130.

[14] F. Guo, H. Rakha, S. Park, Multistate model for travel time reliability. Transp. Res. Record J. Transp. Res. Board 2188 (2010) (2011) 46–54, https://doi.org/10.3141/2188-06.

[15] H. Al-Deek, E. Emam, New methodology for estimating reliability in transportation networks with degraded link capacities. J. Intell. Transp. Syst. Technol. Plann. Oper. 10 (3) (2006) 117–129, https://doi.org/10.1080/15472450600793586.

[16] E.B. Emam, H. Al-Deek, Using real-life dual-loop detector data to develop new methodology for estimating freeway travel time reliability, Transp. Res. Record J. Transp. Res. Board 1959 (2006) (2006) 140–150.

[17] T. Lomax, D. Schrank, S. Turner, R. Margiotta, Selecting Travel Reliability Measures, Texas Transportation Institute, Cambridge Systematics Inc, 2003.

[18] A. Polus, A study of travel time and reliability on arterial routes, Transportation 8 (2) (1979) 141–151.

[19] H. Rakha, I. El-Shawarby, M. Arafeh, Trip travel-time reliability: issues and proposed solutions, J. Intell. Transp. Syst. 14 (4) (2010) 232–250.

[20] H. Tu, J. Van Lint, H. Van Zuylen, Impact of traffic flow on travel time variability of freeway corridors, Transp. Res. Rec. 1993 (2007) (2007) 59–66.

[21] I. Kaparias, M. Bell, H. Belzner, A new measure of travel time reliability for in-vehicle navigation systems, J. Intell. Transp. Syst. 12 (4) (2008) 202–211.

[22] S. Park, H. Rakha, F. Guo, Calibration issues for multistate model of travel time reliability, Transp. Res. Record J. Transp. Res. Board 2188 (2010) (2011) 74–84.

[23] F. Guo, Q. Li, H. Rakha, Multistate travel time reliability models with skewed component distributions. Transp. Res. Record J. Transp. Res. Board 2315 (2012) (2013) 47–53, https://doi.org/10.3141/2315-05.

[24] S. Yang, A. Malik, Y. Wu, Travel time reliability using Hasofer Lind-Rackwitz Fiessler algorithm and kernel density estimation. Transp. Res. Record J. Transp. Res. Board 2442 (2014) 85–95, https://doi.org/10.3141/2442-10.

[25] A. Higatani, T. Kitazawa, J. Tanabe, Y. Suga, R. Sekhar, Y. Asakura, Empirical analysis of travel time reliability measures in Hanshin expressway network, J. Intell. Transp. Syst. 13 (1) (2009) 28–38.

[26] J. Kwon, T. Barkley, R. Hranac, K. Petty, N. Compin, Decomposition of travel time reliability into various sources. Transp. Res. Record J. Transp. Res. Board 2229 (2011) 28–33, https://doi.org/10.3141/2229-04.

[27] H. Wakabayashi, Y. Matsumoto, Comparative study on travel time reliability indexes for highway users and operators. J. Adv. Transp. 46 (4) (2012) 318–339, https://doi.org/10.1002/atr.1194.

[28] M.A. Yazici, C. Kamga, K.C. Mouskos, Analysis of travel time reliability in New York city based on day-of-week and time-of-day periods, Transp. Res. Record J. Transp. Res. Board 2308 (2012) (2012) 83–95.

[29] C. Carrion, D. Levinson, Valuation of travel time reliability from a GPS-based experimental design. Transp. Res. Part C: Emerg. Technol. 35 (2013) 305–323, https://doi.org/10.1016/j.trc.2012.10.010.

[30] A. Sumalee, T. Pan, R. Zhong, N. Uno, N. Indra-Payoong, Dynamic stochastic journey time estimation and reliability analysis using stochastic cell transmission model: algorithm and case studies. Transp. Res. Part C: Emerg. Technol. 35 (2013) 263–285, https://doi.org/10.1016/j.trc.2012.11.003.

[31] S. Susilawati, M.A.P. Taylor, S.V.C. Somenahalli, Distributions of travel time variability on urban roads. J. Adv. Transp. 47 (8) (2013) 720–736, https://doi.org/10.1002/atr.192.

[32] F. Lei, Y. Wang, G. Lu, J. Sun, A travel time reliability model of urban expressways with varying levels of service. Transp. Res. Part C: Emerg. Technol. 48 (2014) 453–467, https://doi.org/10.1016/j.trc.2014.09.019.

[33] A.T. Hojati, Modelling the impact of traffic incidents on travel time reliability. Transp. Res. Part C: Emerg. Technol. 65 (2014) 49–60, https://doi.org/10.14264/uql.2014.492.

[34] S. Yang, C. An, Y.J. Wu, J. Xia, Origin-destination based travel time reliability. Transp. Res. Record J. Transp. Res. Board 2643 (2017), https://doi.org/10.3141/2643-16.

[35] Cambridge Systematics, I, Dowling Associates, I, System Metrics Group, I, Texas Transportation Institute, Cost-Effective Performance Measures for Travel Time Delay, Variation, and Reliability, Transportation Research Board of the National Academies, 2008.

[36] S. Rogers, M. Girolami, A First Course in Machine Learning, Chapman & Hall/CRC, Boca Raton, FL, 2011.

[37] J.B. MacQueen, K means some methods for classification and analysis of multivariate observations, 5th Berkeley Symposium on Mathematical Statistics and Probability 1967, 1(233), 1967, pp. 281–297. https://doi.org/citeulike-article-id:6083430.
[38] S. Frühwirth-Schnatter, Finite Mixture and Markov Switching Models (2006 edition), Springer, New York, 2006.
[39] B. Efron, R.J. Tibshirani, An Introduction to the Bootstrap (1993 edition), Chapman and Hall/CRC, Boca Raton, FL, 1993.
[40] Y. Feng, J. Hourdos, G.A. Davis, Probe vehicle based real-time traffic monitoring on urban roadways, Transp. Res. Part C: Emerg. Technol. 40 (2014) 160–178.

Chapter 5

Urban Travel Behavior Study Based on Data Fusion Model

Meng Li, Mingqiao Zou and Huiping Li
Department of Civil Engineering, Tsinghua University, Beijing, People's Republic of China

Chapter Outline

5.1 INTRODUCTION

This chapter introduces the agent-based traveler behavior model and the relationship of traveler behavior's with VMS board design. Much of this is the author's work, and this chapter can be viewed as an introduction to the agent-based model, which is an extension of traditional max-utility theory. One typical application, based on the traveler behavior model, is to propose a cooperative mechanism and systematic framework of VMS travel guidance and major arterial signal operations [1, 2]. In recent years, many developing countries like China have entered a stage of fast urbanization. As the economy has developed, the number of cars has expanded quickly in China, thus causing serious traffic congestion in many big cities (like Beijing and Shanghai). As a result of this, many traffic management policies and strategies (e.g., vehicle usage restriction, public transport priority) have been introduced to solve the problem, but the result is not as good as expected. One of the most important reasons is that decision makers cannot know exactly how travelers will switch their travel behavior after the policies and strategies are implemented, so the policies and strategies may not meet the actual requirement. Mode choice

Data-Driven Solutions to Transportation Problems. https://doi.org/10.1016/B978-0-12-817026-7.00005-9

and departure time choice are important components of traveler's decision behavior during a trip. At macro level, mode and departure time have a direct bearing on the number and temporal pattern of vehicle trips on urban roadways; at micro level, mode and departure time have great influence on travel time and travel cost. It is necessary to build effective behavior models to predict traveler's departure time shift and travel mode switch, and evaluate the effect of traffic management policies and strategies.

There are many methods to study travel behavior. Traditional theory based on utility maximization assumes substantial rationality and complete information. The logit model derived from the traditional theory is widely used in transportation planning and management. The main expression for the logit model is

$$\ln\frac{p}{1-p}=\beta_0+\beta_1x_1+\cdots+\beta_kx_k \tag{5.1}$$

where β_i is the coefficient of the explaining variable and p is the probability of things happening. The exact probability of things happening is dependent on the actual conditions of travelers like take some certain transportation vehicles. For a multiple-selection situation, the logit model can be expand to

$$\ln\frac{p_i}{p_0}=\beta_{i0}+\beta_{i1}x_1+\cdots+\beta_{ik}x_k \tag{5.2}$$

where p_0 is the basic selection used to make a comparison with other selections.

Some researchers suggest that the assumptions of substantial rationality and complete information are not reasonable and do not conform to the nature of human behavior. A new travel behavior theory has been developed, which considers human being's bounded rationality and incomplete information. Zhang [3] developed the SILK theory, which considers Search, Information, Learning, and Knowledge, for travel decision-making analysis, and has applied the new theory in route choice. Xiong [4] extended the SILK theory into departure time choice. However, there are no applications in mode choice and no research into departure time choice joint with mode choice, especially in developing countries. The details of both theories will be shown in the following literature review section.

This section develops an agent-based joint travel mode and departure time choice model, which considers human being's bounded rationality and incomplete information, and focuses on the traveler's real decision-making process. Travelers are regarded as a group of agents that make decisiosn in the traffic system according to a series of behavior rules. Each agent retrieves spatial, temporal, and travel cost information and accumulates travel experience from every single trip. The agent will update its knowledge through a Bayesian learning process. When a new trip occurs, each agent will make travel mode and departure time choice on the basis of search gain, search cost, search rules, and decision rules.

A joint revealed/stated-preference (SP) survey was conducted in Beijing to calibrate and validate the search gain function, search cost function, search rules, and decision rules. In practice, the agent-based model can be used to analyze the result of travel mode shift and departure time switch for different kinds of policies and strategies that may nudge travelers to change their travel mode and departure time, such as gasoline tax, road pricing, bus and metro fare, and vehicle usage restriction. For example, if we increase the road price during peak period, the proposed agent-based travel behavior mode could be adopted to forecast the peak spreading and mode switch process. This section takes the road network within the Second Ring Road of Beijing to conduct a case study and analyzes the influence on departure time shift and travel mode switch of the different congestion charge policies and the sudden increasing of trip number.

5.2 RESEARCH BACKGROUND

The theory of travel behavior is one of the most important theories in the traffic management and control area. Travel behavior refers to the complicated decision-making process of travelers during a trip, regarding travel mode choice, route choice, departure time choice, destination choice, and so on [5]. Traditional behavior theory assumes that human beings are perfect rational and have perfect information. Before making a decision, individuals can identify all the feasible alternatives. Each alternative consists of a set of attributes that describe that alternative. Travelers judge the level of utility of each alternative based upon those attributes (influenced by the characteristics of the travelers) They then choose the alternative that will maximize their utility. This is called the utility maximization theory [6, 7].

The most widely used travel behavior model based on the maximization theory is the discrete choice model. This is operationalized in the modeling structure by making the choice process a function of both the alternative attributes and the characteristics of the traveler. The study of the discrete choice model in modal share and modal split has a long history. Someone deduced the logit model, which is the typical representative of discrete choice model, on the basis of the research results on theory by McFadden and Ben-Akiva and Lerman, then applied the discrete choice model in a practical project. In order to estimate the probabilities, various forms of logit and probit models have been developed. Most of these models have the problem of independence of irrelevant alternatives (IIA) and are criticized for their inability to consider decision maker's computational capacity and information availability. A large number of researchers have made great efforts to eliminate the IIA property and develops a series of nested logit (NL) models and generalized NL models [7, 8]. In addition, several researches have undertaken the study of mode switching behavior based on the theory of utility maximization [7, 9].

Just the same as the application in mode choice, the multinomial logit departure time choice model also suffers from the problem of irrelevant alternatives

(IAA). In order to avoid IAA, NL models are adopted to analyze correlated departure time intervals [10].

The rational travel behavior theory based on the assumptions of utility maximization and perfect information depicts how travelers should behave, but not how they actually do behave [11]. Some researchers argue that human beings are bounded rationality and have incomplete information. When an individual makes a choice, the information is often incomplete and biased and they will have to expend money and time to search the alternatives and compare them, so they just choose the relatively satisfying alternative, rather than the best one in theory, in most cases [12, 13].

Agent-based modeling (ABM), which considers the bounded rationality and incomplete information of decision makers, is adopted to model travel behavior. ABM focuses on naturalistic (or descriptive) representation of individual behavior and seeks to capture emergent global (or system-wide) patterns resulting from the local interactions and decisions of individual agents [14]. Compared to equation-based modeling (EBM), ABM give more realistic results than EBM in many domains. Previous researchers have conducted some studies in applying ABM to the field of traffic. Arentze et al. [15] proposed an agent-based approach to model activity generation and allocation decisions of individuals and households. Vanek [16] developed an agent-based simulation model of maritime traffic that explicitly models pirate activity and piracy countermeasures. Schelenz [17] applied agent-based simulation for evaluating a bus layout design from passenger's perspective. Yin [18] presented an agent-based travel demand model system for hurricane evacuation simulation. However, these models just emphasize the learning process of individuals and have not relaxed the assumption of rationality in decision-making.

The fuzzy set theory is then adopted to model traveler's decision-making process [19]. Fuzzy set theory is a framework that assumes that decision-making is based on a number of simple "IF, THEN" rules of the form "if … (system perceptions) … then … (preferences toward alternatives) …" [20]. Wehinger et al. [21] developed an agent-based model to study adoption of plug-in hybrid vehicle (PHEV) technology under a variety of scenarios using a set of "if-then" rules. McDonnell and Zellner [22] developed a stylized agent-based model which recreates the decision-making process of choosing a mode though "if-then" rules. Beykaei [23] developed a hierarchical rule-based land-use extraction system. However, the agent-based models that use fuzzy set theory as the behavior rules of agents are relatively infrequent in mode choice research, and no research has been undertaken to study departure time choice jointly with mode choice using this theory. We develop an agent-based joint mode choice and departure time choice model based on the SILK theory. The model is calibrated and validated by using the data from a joint revealed/stated-preference Pad-based and Web-based survey in Beijing. Then the model is applied to analyze the effect of the different congestion charge policies and the sudden increasing of trip numbers within the Second Ring Road of Beijing.

5.3 AGENT-BASED TRAVELER BEHAVIOR MODEL

5.3.1 Travel Behavior Data Collection

In order to calibrate the model, we conducted a joint revealed/SP panel computer-based and web-based survey in Beijing. The survey questionnaire consists of three parts. The first part is a revealed-preference (RP) survey, which contains the respondent's personal attributes and their socio-economic status. The second part is a survey about the respondent's most recent trip; the purpose of this part is to know the current travel behavior of respondents. The third part is the SP experiment, which gives the values of different attributes to respondents according to their answers to the first and second parts of the questionnaire.

5.3.1.1 Revealed-Preference Survey

This part consists of 16 questions, which are designed to establish the socioeconomics and part of the current travel information of respondents. The collected information is displayed as following:

(1) Socioeconomics information of respondents: age, gender, education, occupation, number of people per household, number of children needing shuttling for school per household, income of the respondent per year, income of the whole household per year, and family address.
(2) Travel information: number of vehicles per household, whether respondent has a driving license, whether respondent has a public transit card (can make public transit travel cheaper), average time out of vehicle (e.g., the time for waiting, transferring) for a bus trip, average time out of vehicle for a metro trip, average time out of vehicle for a taxi trip.

5.3.1.2 Last Trip Survey

The purpose of this section is to collect basic information about traveler's commuting and noncommuting trips and the commuting and noncommuting scenario will be given to the respondent with the probability of 50% individually, according to the Beijing Transport Annual Report 2013. The last trip is chosen, on the assumption that respondents will remember the details of this trip so the information will be more reliable. The collected information includes travel mode, departure time, arrival time, expected arrival time, travel distance (TD), and travel cost (fuel cost, parking fee and toll).

5.3.1.3 Stated-Preference Experiment

The last part of this questionnaire is a SP experiment, which can establish the travel behavior switching result of the respondent under different scenarios and polices. There are five kinds of policies and scenarios: (1) increasing of road pricing for car travelers; (2) conducting congestion charge; (3) changing of traffic congestion extent; (4) changing of bus fare; (5) changing of metro fare. The logic design of this is shown in Fig. 5.1.

Firstly, information including travel mode, travel time, and travel cost of the respondent's most recent trip are recorded. Secondly, a policy or scenario is given randomly, then the travel cost and travel time are updated by multiplying a series of random parameters on the basis of the most recent trip information. Thirdly, the respondent is asked to decide whether they would change behavior or not. If the respondent chooses "NO" or "I will cancel the trip," then the survey is stopped; if they choose "YES," then the respondent will be asked to decide what change they would make from alternatives including changing travel mode, changing departure time, and changing route (only for traveler by car and taxi). Lastly, the travel time, travel cost, and travel mode will be updated and displayed to the respondent, then another iteration begins. The interface of the SP experiment is shown in Fig. 5.2.

5.3.1.4 Data Collection

The travel information presented to the respondents in the SP experiment is determined according to the answers collected in Part 1 and Part 2, but the traditional static questionnaire cannot update the travel information dynamically,

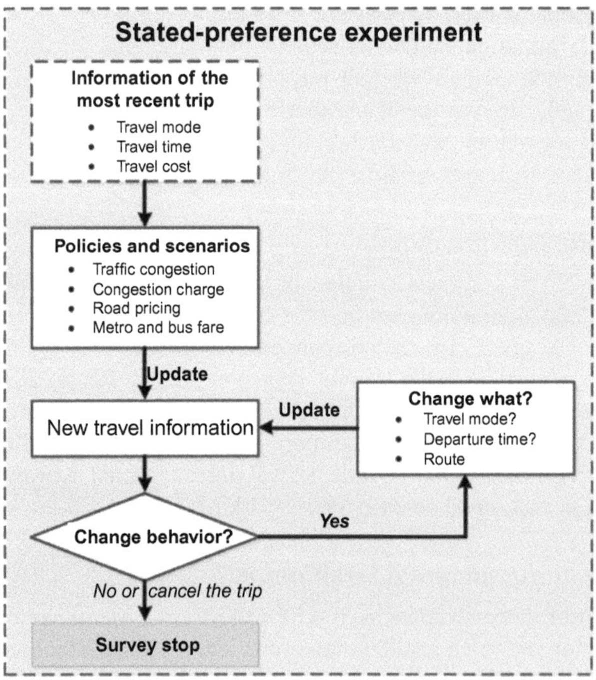

FIG. 5.1 Design of the stated-preference (SP) experiment.

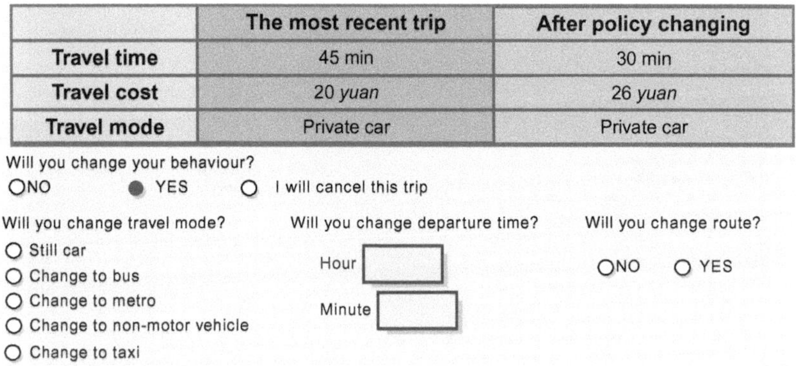

FIG. 5.2 The interface of the SP experiment.

so we have designed a questionnaire which can meet this requirement. The questionnaire was coded in JAVASCRIPT, and JAVA as an APP, to conduct a Pad-based survey, and coded in HTML to conduct a Web-based survey. The survey used simple random sampling methods and the data collection process involved two steps and began on December 10, 2013. First, we sent out the link to the web-based survey to respondents though e-mail, Tencent QQ, Tencent Wechat, and Fetion. In addition, the link was displayed on the website of Renren and Sina Weibo. We collected a total of 162 effective samples though the Web-based survey in two weeks.

After comparing the gender ratio and age distribution of the 162 samples with the Sixth Nationwide Population Census Data of Beijing, we conducted the supplement survey by using the Pad-based questionnaire on December 25, 2013; this could be targeted at specific samples, thus making the whole survey data more reliable. Another 81 effective samples were collected though the Pad-based survey, giving 241 effective samples in total. The comparison of the gender ratio between the survey data and the Sixth Nationwide Population Census Data is shown in Fig. 5.3 and Fig. 5.4.

As shown in Fig. 5.5, the departure time distribution achieved in this survey is slightly different from the result of the Sixth Nationwide Population Census Data. The morning peak hour is 7:00–9:00 am, the evening peak hour is 17:00–19:00.

Fig. 5.6 shows that the mode share of public transit (Metro, 12% and Bus, 28%) in Beijing is nearly the same as the driving mode (car, 40%). The total share of car, bus and metro is about 80%.

5.3.2 Model Development

5.3.2.1 Model Framework

According to the typical structure of an agent, we develop the framework of the agent-based choice model, as shown in Fig. 5.7 Individuals accumulate travel

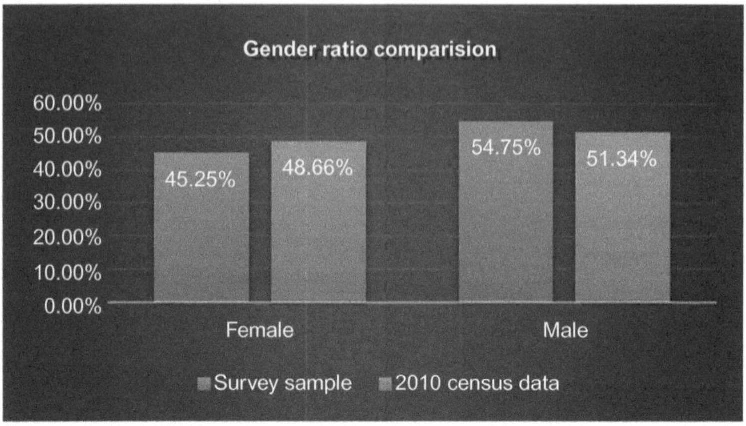

FIG. 5.3 Comparison of the gender ratio.

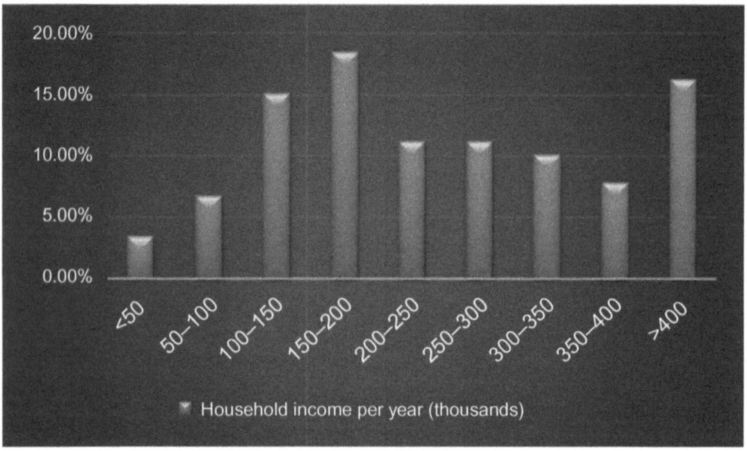

FIG. 5.4 Household income distribution.

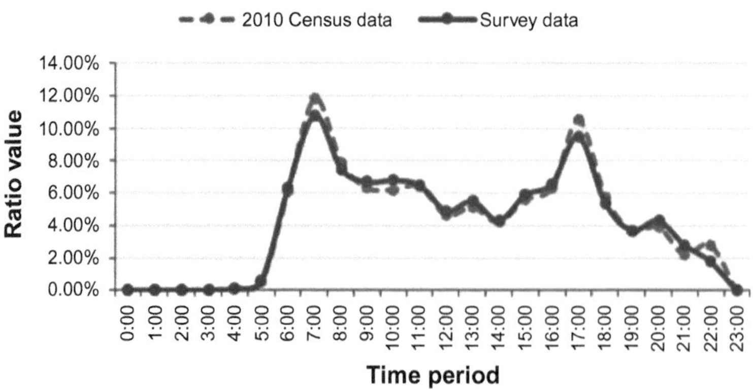

FIG. 5.5 Departure time distribution.

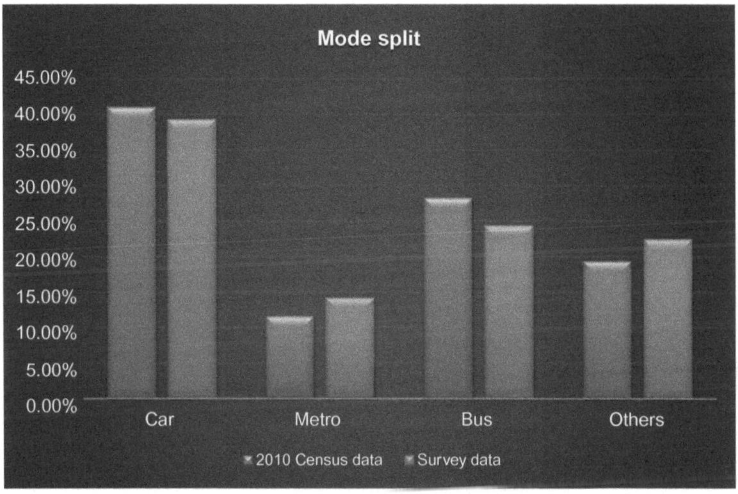

FIG. 5.6 Mode split.

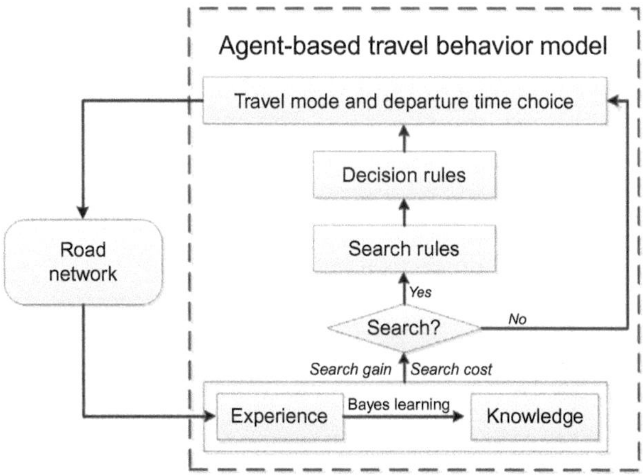

FIG. 5.7 Framework of the agent-based choice model.

experience though the performance information of the road network and other conditions, such as traffic management policies and strategies. Each traveler will achieve spatial and temporal knowledge through a Bayesian learning process.

Travelers typically face two types of pretrip choices: mode selection and departure time selection. For a repeating trip, travelers might choose to keep the current mode and departure time or search for new options. In order to model when a traveler will begin and stop searching, the theory of search gain and search cost with imperfect information is adopted. The search gain is the individual's subjective benefits from an additional search. For example, if the current travel mode is car and the alternative mode is metro, then possible travel cost savings predicted by the traveler is the search gain. The search cost is the time, money and energy paid for an additional search by the traveler.

If the search cost is bigger than the search gain, travelers will decide not to search, then repetitive learned behavior or habitual behavior is executed. Otherwise, travelers will search for alternative modes according to a series of heuristic search rules. When a feasible alternative mode or departure time is searched out, a group of decision rules are adopted by travelers to decide whether to switch to the new mode and departure time or not.

5.3.2.2 *Knowledge Learning Process*

An individual's spatial knowledge for travel mode choice is based on experienced utility from previous learning and trials. Assume that there are I kinds of travel modes (or departure times) based on prior perception, and that mode mi with utility u_i has been experienced n_i times by a particular individual. Therefore, the individual's knowledge about travel modes can be quantified as a vector $K(n_1, \ldots n_i, \ldots n_I)$. According to Bayesian learning rules, the perceived weights of past observations are the same. Let vector $P(p_1, \ldots p_i, \ldots p_I)$ represent an individual's subjective beliefs, where pi is the subjective probability that an additional search would lead to an alternative travel mode with utility u_i.

We assume that individual's prior beliefs follow a Dirichlet distribution to establish a quantitative relationship between knowledge K and beliefs P. Since the Dirichlet is the conjugate prior of the multinomial distribution, the posterior beliefs will also be a Dirichlet distribution. This assumption is equivalent to Eq. (5.3), where N denotes the total number of observations ($N = sum(n_i)$).

$$P_i = \frac{n_i}{N} \tag{5.3}$$

The knowledge learning process for departure time can be seen in Ref. [24].

5.3.2.3 *Search Gain and Search Cost*

The decision to search for a new alternative travel mode is based on the relationship between subjective search gain and search cost. We assume that an individual's utility associated with the current travel mode is u. The subjective search gain (g) is the expected utility improvement from an additional search:

$$g = \sum_{i(u_i > u)} p_i(u_i - u) \tag{5.4}$$

Define the theoretically maximum utility as u. Then after N searches, the probability of finding a travel mode with utility u^* is $\frac{1}{N+1}$. Thus we can simplify Eq. (5.4) as follows:

$$g = (u^* - u_{\max})/(N+1) \tag{5.5}$$

Furthermore, perceived search cost needs to be estimated and then compared with subject search gain in order to initiate the search process. The perceived search cost is assumed to be constant for the same individual throughout the process. If an individual stops searching after n rounds of search, the perceived search cost must be lower than the subjective search gain after the $(n-1)$th search so that the nth search is meaningful, and must be higher than the expected search gain after the nth search so that the $(n-1)$th search does not occur. The lower bound c^- and upper bound c^+ can be calculated though Eqs. (5.6), (5.7), then we can estimate the search cost as shown in Eq. (5.8).

$$c^- = g_n = (u^* - u_{\max,n})/(n+1) \tag{5.6}$$

$$c^+ = g_{n-1} = (u^* - u_{\max,n-1})/n \tag{5.7}$$

$$c = (c^- + c^+)/2 \tag{5.8}$$

The utility function adopted in this paper is a linear function, which contains five variables: travel time, travel cost, a dummy variable, TDE, and TDL (defined in next part). The utility function is estimated with the survey data in Part 1 and Part 2 of the questionnaire. Because the search times of each respondent are recorded in the SP experiment, the search cost distribution is easily achieved.

5.3.2.4 Search Rules

There are many ways to describe the human knowledge storage and application process. Production (if-then) rules are selected to represent the search process of human beings for several reasons:

(1) They have been shown to be capable of replicating various types of human heuristic decision-making processes in previous studies on expert systems.
(2) They can be implemented to predict search results with minimum computational resources, which is important for mode and departure time choice models often involving millions of independent decisions agents.
(3) Production rules can express the knowledge of human beings naturally and conform to the cognition process.
(4) Models based on production rules are quick to calculate.

The data set collected from the Pad-based and Web-based travel behavioral survey in Beijing was used to derive the search rules. Because mode switching between metro, bus and nonmotorized (walk and bike) is not so meaningful for traffic policies and strategies, we just study the mode switch between car and other modes. In addition, taxi and private car have a similar influence on the traffic system, so we regard them as the same mode: car. After analyzing the survey data, we find out that only 0.85% of travelers will switch to car, and there are few switches between car and nonmotorized modes, so we just study the search process from car to bus and metro. The search process contains two dimensions: travel model and departure time search. The variables used in the search rules include: age (AGE), personal income per year (PI), household income per year (HI), TD, outside metro time (OMT), outside bus time (OBT), DTL, DTE, D-TIME, and D-COST. OMT (OBT) means the time travelers spend going to the metro (bus) station, waiting, and transferring. DTL and DTE means the difference value between preferred arrival time (PAT) and actual arrive time (AAT). D-TIME means the increment of travel time, D-COST means the increment of travel cost. See Eq. 5.9:

$$\begin{gathered} DTL = AAT - PAR \quad AAT \geq PAT \\ DTE = PAT - AAT \quad PAT > AAT \\ D-COST = COST^1 - COST^0 \\ D-TIME = TIME^1 - TIME^0 \end{gathered} \tag{5.9}$$

where AAT is the actual arrival time, PAT is the preferred arrival time, $TIME^1$ and $COST^1$ are the travel time and travel time after changes, $TIME^1$ and $COST^1$are the values before changes. Machine learning algorithm adopted to derive the search rules [25]. The search rule sets for travel mode and departure time are shown below.

The Current Travel Mode Is Car	
Search bus, if	
Rule 1	[AGE > 45 AND D-COST > 5 yuan AND PI < 50,000 yuan] (12.0/1.0)
Rule 2	[AGE > 37 AND D-COST > 40 yuan AND OMT > 30 min] (15.0/1.0)
Rule 3	[150,000 yuan < HI ≤ 200,000 yuan AND 100,000 yuan < PI ≤150,000 yuan AND D-COST > 10 yuan AND OMT > 20 min] (3.0)
Rule 4	[OMT > 30 min AND 10 < D-COST ≤ 45 yuan] (5.0/1.0)
Search metro, if	
Rule 5	[D-TIME > 25 min AND 50,000 yuan < PI ≤100,000 yuan] (10.0)
Rule 6	[D-COST > 40 yuan AND 300,000 yuan < PI ≤350,000 yuan AND AGE > 28] (7.0/1.0)
Rule 7	[D-TIME > 25 min AND HI > 400,000 yuan AND OMT ≤ 30 min AND D-COST > 0] (8.0/1.0).
Rule 8	[50,000 yuan < PI ≤ 100,000 yuan AND OBT > 10 min AND D-COST > 20 yuan] (14.0/2.0)
Rule 9	[150,000 < PI ≤ 200,000 yuan AND D-COST > 5 yuan AND OBT > 10 min] (10.0/1.0)
Rule 10	[AGE ≤ 28 AND OBT ≤ 30 min AND 150,000 yuan < HI ≤ 200,000 yuan AND D-COST > 40 yuan] (5.0)
Rule 11	[HI ≤ 50,000 yuan AND DTE < 15 min AND AGE ≤ 26 AND D-COST > 0] (6.0/1.0)
Rule 12	21 < AGE ≤ 28 AND 200,000 yuan < HI ≤ 250,000 yuan AND D-TIME > 10 min] (5.0/1.0)
Rule 13	[35 yuan < D-COST ≤ 60 yuan AND OBT > 30 min AND OMT ≤ 20 min (8.0/1.0)

Rule 14	Otherwise, continue to use car and begin to search departure time (108/20)
Rule 15	Search 0–20 min earlier, if: [0 < DTL ≤ 20 min] (4.0)
Rule 16	Search 20–30 min earlier, if: [10 min < DTL ≤ 20 min AND TD > 20 km] (16.0/4.0)
Rule 17	Search 30–40 min earlier, if: [20 min < DTL ≤ 45 min AND TD ≤ 15 km] (15.0/3.0)
Rule 18	Search 40–60 min earlier, if: [30 min < DTL ≤ 50 min AND TD > 15 km] (20.0/3.0)
Rule 19	Search 60 + min earlier, if: [DTL > 50 min] (24.0/4.0)
Rule 20	Search 0–20 min later, if: [20 min < DTE ≤ 40 min AND TD ≤ 20 km) (34.0/6.0)
Rule 21	Search 20–40 min later if: [30 min < DTE ≤ 60 min AND TD ≤ 30 km] (43.0/5.0)
Rule 22	Search 40+ min later, if: [DTE > 60 min] (15.0/2.0)
Rule 23	Otherwise, keep the current departure time as the alternative of the next trip (42.0/0)

There are 23 rules in the search rule sets that are just for car travelers. Rule 1 means that if an individual's age is over 45, and the travel cost has an increase of 5 yuan (i.e., congestion charge), and PI is below 50,000 yuan per year, that individual will search for bus as the alternative travel mode. It is reasonable, because bus is cheaper than metro in Beijing. When a traveler will not search for an alternative mode, according to the rules from 1 to 13, then he or she will continue to use car as the travel mode and begin to search for alternative departure times according to rules from 15 to 22.

5.3.2.5 Decision Rules

After each round of searching, a new travel mode and departure time is identified by the search rules. The traveler then should decide whether to stick to the current travel mode and departure time or switch to the new mode and departure time according to decision rules. Similar to search process, if-then rules are selected to represent the decision-making process of human beings. The machine learning algorithm, JRip [25a], is adopted to derive the decision rules. The decision variables used in rule set include: AGE, PURPOSE, GENDER, DTIME, DCOST, INCOME.

$$\begin{aligned} \Delta TIME &= \left(TT^1 - TT^0\right)/TT^0 \\ \Delta COST &= \left(C^1 - C^0\right)/C^0 \end{aligned} \tag{5.10}$$

where TT^1 is the travel time with the alternative travel mode or departure time, TT^0 is the travel time before switch. C^1 is the travel cost with the alternative travel mode or departure time and C^0 is the travel cost before switch. The decision rules are presented below.

Decision Rules	
The Current Mode is Car, Switch to the alternative mode, if	
Rule 1	[$\Delta TIME \le -0.066667$ AND AGE ≤ 30 AND PURPOSE = noncommute] (51.0/6.0)
Rule 2	[$\Delta TIME \le 0.105263$ AND $\Delta COST \le -0.022727$ AND age ≤ 30 AND GENDER = female] (13.0/1.0)
Rule 3	[$\Delta TIME \le -0.1$] (15.0/4.0)
Rule 4	[INCOME $\ge$400,000 AND AGE ≤ 33] (19.0/4.0)
Rule 5	[$-0.287129 \le \Delta TIME \le -0.066667$ AND $\Delta COST \le -0.75$ AND AGE $\ge$32] (8.0/0.0)
Rule 6	**Otherwise**, continue to use the current mode (523.0/45.0)
The Current Mode is Car, Switch to the alternative departure time, if	
Rule 7	[$\Delta COST \le 0$ AND $\Delta TIME \le -0.1$] (125.0/15.0)
Rule 8	**Otherwise**, continue to use the current departure time

Eight rules are derived from the survey data. Rule 1 indicates that if the travel time decreases by >6.67%, and the age is <30 years old, and the trip is noncommute travel, then the traveler will switch to the new mode. Rule 7 shows that if the travel cost can be reduced, and the travel time can be reduced by at least 10%, then the traveler will switch to the new departure time.

5.3.3 Policy and Scenario Analysis

5.3.3.1 Analysis Framework

The agent-based joint travel mode and departure time choice model can be combined with macro –/meso/micro-scopic network traffic models to analyze the modal shift and peak spreading process under different policies and scenarios. As shown in Fig. 5.8, at first, an initial OD matrix and an initial dynamic network user equilibrium are given. When policies and scenarios (e.g., gasoline tax, congestion charge, and sudden increase in demand) are applied to the network, the equilibrium state is disturbed and travelers will conduct modal shift and departure time switch following the agent-based travel behavior model until behavior equilibrium (BE) is obtained. The corresponding simulation network in our research is displayed in Fig. 5.9.

5.3.3.2 Analysis of Demand Increase

We assume that there are 24,000 agents on the network at first, and their initial mode is all by car. We increase the demand gradually to observe the changing process of average speed, modal changing rate (the percentage of traveler switch to bus and metro), and departure time.

As shown in Fig. 5.10, when the total number of travelers is below 80,000, the changing rate is almost constant, at a value of only about 1.50%; the average speed (considering outside vehicle time) of car, bus, and metro also remain

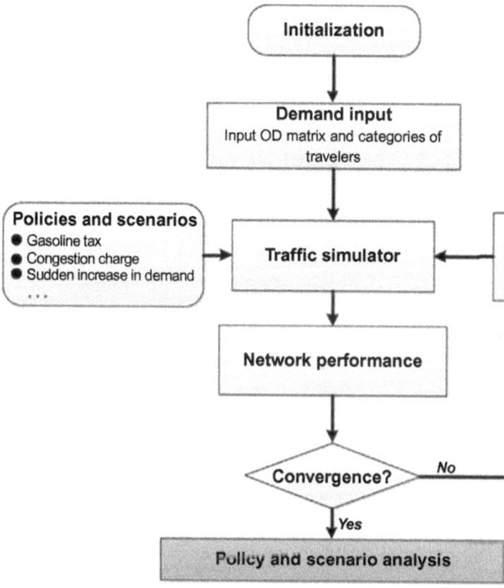

FIG. 5.8 Policy and scenario analysis framework.

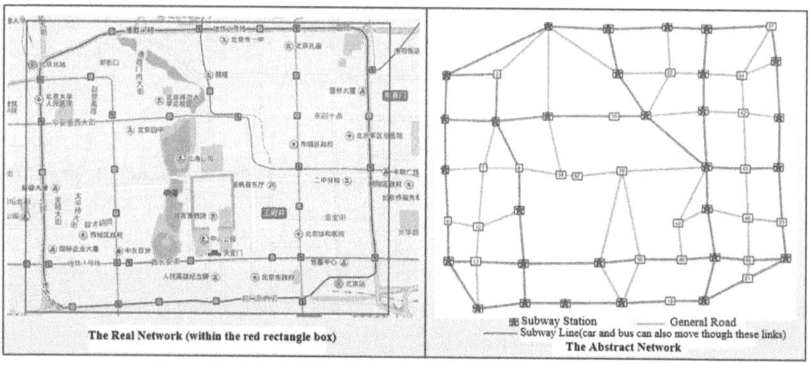

FIG. 5.9 Simulation network (2nd ring road of Beijing).

unchanged. When total number exceeds 80,000, the changing rate ascends gradually along with the increase of travelers; the average speed of car and bus decrease at the same time, while the average of metro almost remains the same because of the low sensitivity to volume. When the total number is 200,000, the changing rate is 8.29%, among which 5.73% of travelers shift to metro, 2.56% shift to bus. The effects of peak spreading are shown in Fig. 5.11.

Fig. 5.10 shows that travelers who do not shift mode will switch their departure time due to the increased demand to reduce congestion; the greater the increase in demand, the more significant the peak spreading phenomenon. When the total number of travelers is 200,000, there are about 23.31% switching their departure time from 8:00 am to other alternatives.

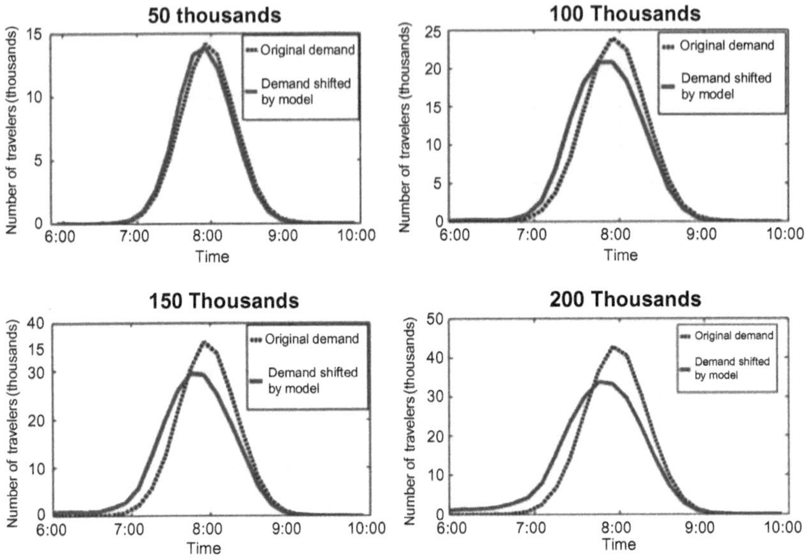

FIG. 5.10 Congestion charges scenarios (I).

5.4 BEHAVIOR MODEL IN COOPERATION OF VMS AND TRAFFIC SIGNAL

5.4.1 Drivers' Diversion Model

In developing countries, VMS are widely used in urban road network. These VMS panels allow the display of information on the road network "congestion scale," represented with different colors. The "congestion scale" is a classification of average speed calculated by travel time (i.e., red means congested road segment; yellow means heavy traffic volume segment; and green means low-traffic segment). In detail, for major arterials, red represents the real-time speed lower than 20 km/h; yellow represents 20–40 km/h; and green represents higher speed than 40 km/h. (see Fig. 5.12) Previous research focuses on drivers' response to typical text messages while neglecting the graphical information on VMS panels.

To solve existing problems of drivers' responses to graphical information, we carried out a SP survey, which is coded into pad computers. Considering road segments with different colors and detour distances that drivers might face in the urban area, various variables including distances of pretrip path (AB) and detour route (ACDB), diversion advice, and colors were changed randomly.

Multiple factors vis-à-vis VMS have been considered in the SP survey and the collected information can be classified in the following categories:

(1) Personal information: age, gender, and salary; and the personal attributes which define drivers' characteristics have been identified in Table 5.1.
(2) Preference information: (a) a driver's acceptability of VMS; (b) a driver's response to information on VMS.

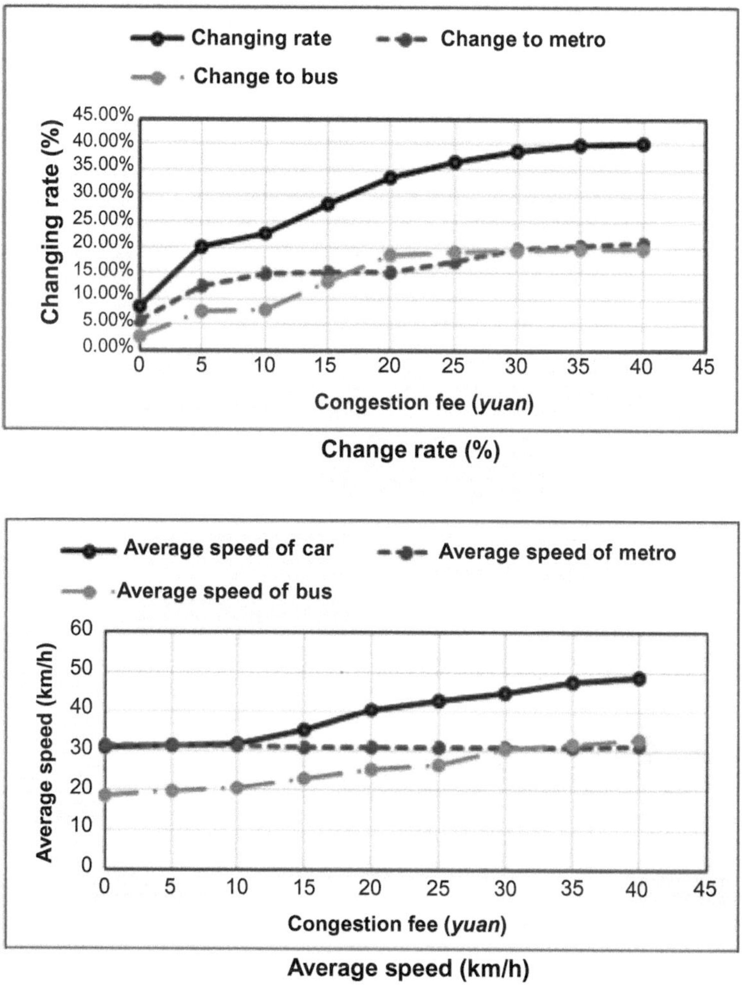

FIG. 5.11 Congestion charges scenarios (II).

(3) Dynamic information on VMS panels: (a) distances: original route and alternative routes; (b) colors: yellow and red segment ratios; (c) diversion advice.

By analysis of collected behavioral data, we managed to find a set of significant factors. A binary logit model is presented in Table 5.2. The estimated coefficients of the variables are all significant at 95% confidence level. The likelihood of diversion has been positively affected by certain factors, for example, age, length of routine route l_{AB}, diversion advice, red segment ratio on the routine route RLR_0 and passively affected by length of diversion route l_{ACD}, and red segment ratio on the diversion route RLR_1. In addition, the utility function is formulated as:

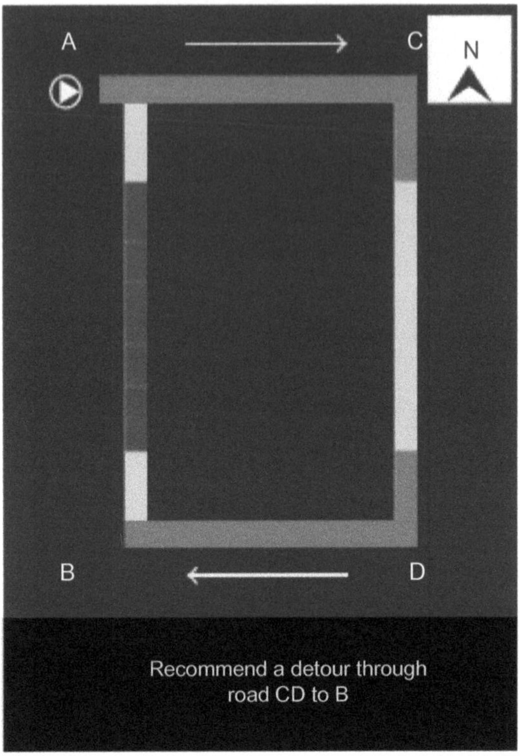

FIG. 5.12 An illustration of a VMS panel.

TABLE 5.1 Summary of Selected Personal Attributes

Personal Attributes		Percentage (%)
Gender	Male	44.94
	Female	55.06
Age	18–29	39.33
	30–49	58.43
	50–69	2.25
Personal salary	<50,000	15.73
	50,001–100,000	21.35
	100,001–150,000	20.22
	150,001–200,000	11.24
	200,001–250,000	16.85
	250,001–300,000	1.12
	300,001–350,000	4.49
	350,001–400,000	1.12
	>400,000	7.87

TABLE 5.2 Binary Logit Model for Drivers' Responses to VMS

Variables	Coefficient	Estimates	Standard Error	Significance
Age	β_1	0.044	0.017	0.010
l_{AB}	β_2	0.965	0.096	0.001
l_{ACB}	β_3	−0.439	0.076	0.001
RLR_0	β_4	3.510	0.606	0.000
RLR_1	β_5	−6.240	1.371	0.000
Advice	β_6	0.431	0.211	0.041
Constant	β_0	−3.032	1.314	0.021

$$\begin{aligned} V_n(C) &= V_n(1) - V_n(0) \\ &= \exp(\beta_0 + \beta_1 \cdot \text{Age} + \beta_2 \cdot l_{AB} + \beta_3 \cdot l_{ACD} + \beta_4 \cdot RLR_0 \\ &\quad + \beta_5 \cdot RLR_1 + \beta_6 \cdot \text{Advice} + \text{const}) \end{aligned} \tag{5.11}$$

The probability that individual n chooses alternative route can be obtained in the following equation:

$$P_n(C) = \frac{V_n(C)}{1 + V_n(C)} \tag{5.12}$$

5.4.2 Cooperative Mechanism of VMS and TSC

5.4.2.1 Cooperative Mechanism

Based on the previously proposed logit model, we found that the red segment ratios of an arterial displayed on VMS have a profound impact on the behavior of en-route diversion decisions. Consequently, it will be a reasonable and practical choice to take the dynamic graph information as an active travel guidance tool (i.e., to display arterial with colors predicted by a reasonable and desirable method), instead of displaying current traffic status passively. With the collaboration of traffic signal control, the new VMS strategy method can not only distribute traffic flow in desirable proportions but also provide traffic status related more closely to traveler's driving experience.

The important improvement of the active VMS strategy is achieved by two preconditions: first, we allow the cooperation of VMS and TSC systems instead of using them in isolation; second, and more importantly, we can find a desirable solution by joint optimization in the short term, and the solution has to be consistent with drivers' driving experience for a sustainable effectiveness in a long-term use. This approach strikes a very nice balance between accessibility and effectiveness.

5.4.2.2 Simulation-Based Method for the Two-Stage Nested Optimization Problem

We provide a simulation-based-optimization (SBO) method for the two-stage nested optimization problem. To better illustrate the cooperative mechanism, we propose a systematic framework in Fig. 5.13. In view of the complexity of network traffic flow and interactions between different strategies, SBO methods have been introduced to solve a few engineering optimization problems, such as a decision support tool for mitigating traffic congestion and regional signal timing strategies optimization [26].

With multiple decision variables in the two-stage optimization problem, the interdependence between VMS and traffic signal control is hard to evaluate by a traditional optimizer. In addition, a microscopic traffic simulator can give a relatively accurate estimation of flow distribution, average speeds, and interactions of vehicles.

Therefore, we adopt the SBO method to solve the aforementioned optimization problem. In Fig. 5.13, the strategy module concerns operational strategies: VMS travel guidance and traffic signal control. Based on the inputs (e.g., network, demand, and strategies), the simulation module in TRNSIMS can produce detailed outputs including accurate traffic conditions and traffic flow assignment with the driver diversion module. Within the framework, the optimization module is designed to solve the first-stage optimization problem by using genetic algorithms (GAs); the second-stage optimization problem is solved by a dynamic programming method. The SBO framework aims to obtain feasible optimal strategies that minimize total travel time of all travelers.

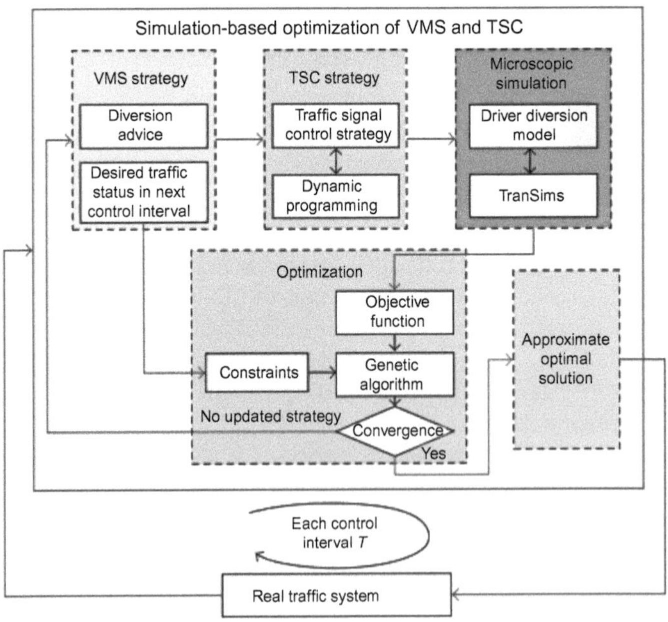

FIG. 5.13 An SBO framework for the VGSC problem.

5.4.3 Applications

5.4.3.1 Topology

The proposed methodology is applied to the actual road network of Tianjin Binhai Hi-Tech Industrial Development Park (THIP, China). This site is located 3 km west of the Tianjin urban area, with a 2 km^2 core area. As a connection to the Tianjin municipal area, the road network in this core area is almost filled with heavy traffic flow during rush hours on working days. In this area, >80% of traffic is due to daily commuters who work in THIP. Moreover, traffic management rules and traffic signals are properly maintained, but daily traffic congestions still trouble travelers. Therefore, the road network of THIP. is selected for the proposed study (as shown in Fig. 5.14). It has 30 nodes and 46 unidirectional links. Among them, there are two arterials: Fukang Road (Path 1, 1.97 km, in Fig. 5.15) and Yingshui Road (Path 2, 3.9 km, in Fig. 5.15), which carry the greatest traffic volume from the Tianjin urban area to THIP during morning peak hours (from 8:00 to 10:00 am). Besides the links and nodes inside this area, we also set some external zones (blue trapezoids in Fig. 5.15) on the boundary of the road network, which are connected with external links to the major traffic flow.

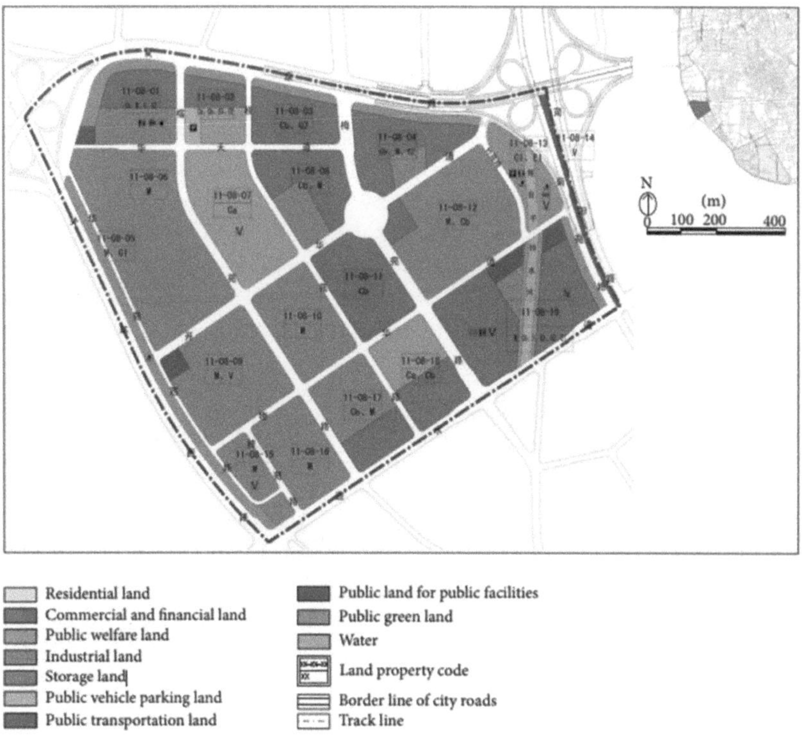

FIG. 5.14 Map of THIP with land use.

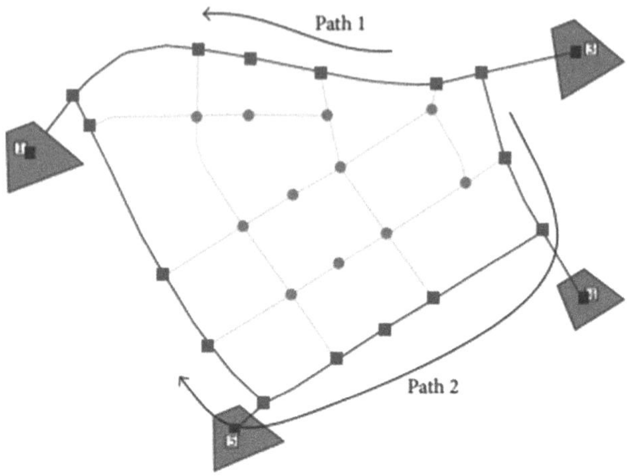

FIG. 5.15 Road network topology of THIP.

5.4.3.2 Simulation Results and Analysis

According to the size of the study network, we set the control interval as 5 min. The 12 optimum solutions can be obtained during 8:00–9:00 in the morning peak time. During each interval, seven variables are optimized in the first-stage GA optimization, including the desired yellow ratios on the two arterials, the cycle times and green times for each arterial, and diversion advice. Moreover, the offsets of 10 signalized intersections on the two arterials are optimized by the second-stage dynamic programing.

By the multiuse of the three control methods under traffic demand during the morning peak time, we present the mean values of 20 groups. The comparison of improvement of traffic conditions by different strategies is shown in Table 5.3. The improvement by adopting the active VMS strategy with fixed signal control is 10.2%, while the cooperation of both strategies reaches 26.3%.

The convergence process of each control interval is explained further in Fig. 5.16. There exists an obvious trend of convergence when the generation

TABLE 5.3 Comparison of Minimum Values of Objective Function

Control Method	Index	
	Objective Function Values	*Improvement*
Pretimed signal control without VMS	84,756.77	–
Pretimed signal control with VMS strategy	76,118.13	10.2%
The cooperative strategy of VMS and arterial signal control	62,391.83	26.3%

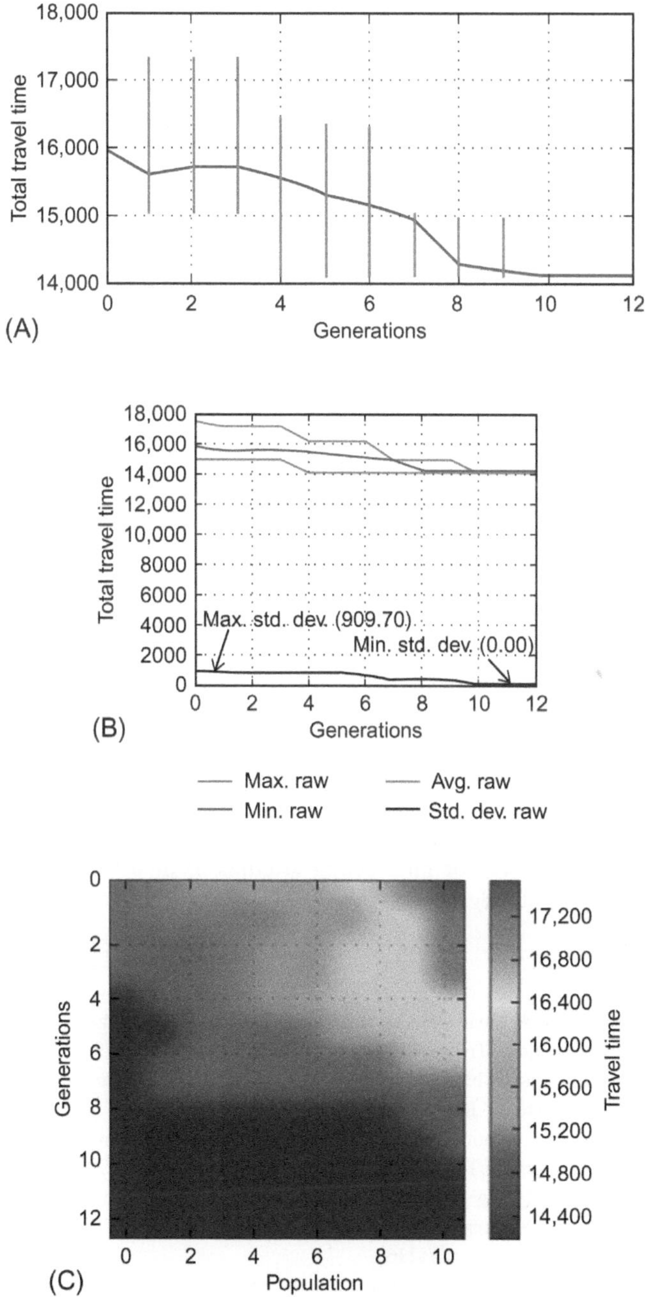

FIG. 5.16 Convergence process of the genetic algorithm: (A) The evolution process, (B) the standard deviation of population in generations, and (C) total travel time of population along generations.

increase as shown in Fig. 5.16A. The maximum number of generations is set to 20; the mutation rate is 0.02; and the crossover rate is 0.9. When the optimization comes to the 10th generation, the standard deviation is close to zero, which satisfies the convergence criteria in GA in Fig. 5.16B. It can be explicitly shown that the objective function values of each individual from the 10th generation are kept the same as shown in Fig. 5.16C.

5.5 CONCLUSIONS

This chapter developed an agent-based based choice model for travel mode and departure time according to the behavior survey conducted in Beijing, China. This research is an extension of the traditional max-utility theory and focuses on the issues in developing countries, where travel behavior is different from that in developed countries. Different from the traditional utility maximization based model, this model considers the bounded rationality of human beings and focuses on the knowledge learning, search, and decision-making process of agents.

One interesting application is used in the cooperative mechanism of VMS travel guidance and arterial signal control for urban traffic management. In this mechanism, traffic control parameters and VMS parameters are optimized together. Particularly for arterial signal control, a two-stage nested optimization problem is formulated. To find the optimal solution, we apply a simulation-based optimization framework for solving this two-stage nested optimization problem. In this cooperative strategy, we allow the communication between VMS and TSC systems to find a desirable and feasible solution during each control interval. In the future, we will focus on testing the proposed method for more complex networks of intersecting arterials to investigate effectiveness of the cooperative strategy.

REFERENCES

[1] D. Brownstone, K.A. Small, Valuing time and reliability: assessing the evidence from road pricing demonstrations, Transp. Res. A Policy Pract. 39 (4) (2005) 279–293.

[2] M. Zou, et al., An agent-based choice model for travel mode and departure time and its case study in Beijing, Transp. Res. C 64 (2016) 133–147.

[3] L. Zhang, Search, Information, Learning, and Knowledge in Travel Decision-Making: A Positive Approach for Travel Behavior and Demand Analysis, ProQuest, 2006.

[4] C. Xiong, All About Congestion: Modeling Departure Time Dynamics and Its Integration With Traffic Models, Digital Repository at the the University of Maryland (DRUM), (2011). https://drum.lib.umd.edu/handle/1903/12360.

[5] C. Wen, F.S. Koppelman, A conceptual and methdological framework for the generation of activity-travel patterns, Transportation 27 (1) (2000) 5–23.

[6] J. Von Neumann, O. Morgenstern, Theory of Games and Economic Behavior, Princeton University Press, Princeton, NJ, 2007.

[7] T. Wang, C. Chen, Attitudes, mode switching behavior, and the built environment: a longitudinal study in the Puget Sound Region, Transp. Res. A Policy Pract. 46 (10) (2012) 1594–1607.

[8] C. Wen, F.S. Koppelman, The generalized nested logit model, Transp. Res. B Methodol. 35 (7) (2001) 627–641.

[9] K.K. Srinivasan, H.S. Mahmassani, Analyzing heterogeneity and unobserved structural effects in route-switching behavior under ATIS: a dynamic kernel logit formulation, Transp. Res. B Methodol. 37 (9) (2003) 793–814.

[10] M. Ben-Akiva, M. Bierlaire, Discrete Choice Models With Applications to Departure Time and Route Choice, Springer, 2003, pp. 7–37.

[11] L. Zhang, Agent-Based Behavioral Model of Spatial Learning and Route Choice, TRB (Transportation Research Board), 2006.

[12] A. Acquisti, J. Grossklags, Privacy and rationality in individual decision making, IEEE Secur. Priv. 3 (1) (2005) 26–33.

[13] G. Wei, GRA method for multiple attribute decision making with incomplete weight information in intuitionistic fuzzy setting, Knowl.-Based Syst. 23 (3) (2010) 243–247.

[14] M. Zou, et al., Dynamic transportation planning and operations: concept, framework and applications in China, Procedia Soc. Behav. Sci. 96 (2013) 2332–2343.

[15] T.A. Arentze, D. Ettema, H.J. Timmermans, Incorporating time and income constraints in dynamic agent-based models of activity generation and time use: approach and illustration, Transp. Res. C 18 (1) (2010) 71–83.

[16] O. Vaněk, et al., Agent-based model of maritime traffic in piracy-affected waters, Transp. Res. C 36 (2013) 157–176.

[17] T. Schelenz, et al., Application of agent based simulation for evaluating a bus layout design from passengers' perspective, Transp. Res. C 43 (2014) 222–229.

[18] W. Yin, et al., An agent-based modeling system for travel demand simulation for hurricane evacuation, Transp. Res. C 42 (2014) 44–59.

[19] P.C. Vythoulkas, H.N. Koutsopoulos, Modeling discrete choice behavior using concepts from fuzzy set theory, approximate reasoning and neural networks, Transp. Res. C 11 (1) (2003) 51–73.

[20] H.J. Zimmermann, Fuzzy set theory, Wiley Interdiscip. Rev. 2 (3) (2010) 317–332.

[21] L.A. Wehinger, M.D. Galus, G. Andersson, Agent-Based Simulator for the German Electricity Wholesale Market Including Wind Power Generation and Widescale PHEV Adoption, IEEE, Zürich, Switzerland, 2010.

[22] S. McDonnell, M. Zellner, Exploring the effectiveness of bus rapid transit a prototype agent-based model of commuting behavior, Transp. Policy 18 (6) (2011) 825–835.

[23] S.A. Beykaei, et al., A hierarchical rule-based land use extraction system using geographic and remotely sensed data: a case study for residential uses, Transp. Res. C 47 (2014) 155–167.

[24] L. Zhang, C. Xiong, How do Travelers Actually Adjust Departure Times: A Positive Model of Peak Spreading Dynamics, TRB (Tansportation Research Board), 2012.

[25] W. Klösgen, J.M. Zytkow, Handbook of Data Mining and Knowledge Discovery, Oxford University Press, Oxford, England, 2002.

[25a] I.H. Witten, E. Frank, Data Mining: Practical Machine Learning Tools and Techniques, second ed., Morgan Kaufmann, San Francisco, 2005.

[26] M. Li, H. Jiang, Z. Zhang, W. Ni, P. Zhang, J. Song, A simulation-based framework for the cooperation of vms travel guidance and traffic signal control, Math. Probl. Eng. 2014 (1) (2014) 1–13.

Chapter 6

Urban Travel Mobility Exploring With Large-Scale Trajectory Data

Jinjun Tang
School of Traffic & Transportation Engineering, Central South University, Changsha, China

Chapter Outline

6.1 INTRODUCTION

Human travel behaviors are affected by a number of factors such as spatial structure of a city, land use, and road networks. Understanding the regularity and characteristics of human mobility is of major importance to city and

Data-Driven Solutions to Transportation Problems. https://doi.org/10.1016/B978-0-12-817026-7.00006-0

transportation planning. As a questionnaire based approach is constrained by lack of data, it is difficult to use this traditional method to explore human mobility deeply and accurately. The fast development of information and communication technology makes it possible to understand travel behaviors of people by providing large-scale and granular data, recording individual information chronologically. Various datasets, including wireless network traces [1], GPS traces from probe vehicle data [2–5], cell phone [6–15], and banking notes [16], are collected to study spatial-temporal feature of human movement.

Jiang et al. [2] analyzed the human mobility pattern from over 72,000 people's moving trajectories, collected from 50 taxicabs during a 6-month period. Zheng et al. [3] proposed a graph-based postprocessing algorithm to infer human movement modes from GPS trajectories data. In order to analyze the accessibility, Li et al. [5] introduced a new dynamic accessibility measure based on real-time travel speed extracted from probe vehicle data. Csáji et al. [6] used principal component analysis to reveal the relation between features of human behavior and their geographical location from cell phone datasets. Sun et al. [7] also applied principal component analysis to discover the urban dynamics based on cell phone location information. Kang et al. [8] presented the distribution of human urban travel following the exponential law, in which the exponents were affected by city size and shape, and used Monte Carlo simulation to verify the relation between intra-urban human mobility and urban. By constructing a cell phone network, Hidalgo et al. [9] defined the persistence of ties to explore dynamics of human mobility. González et al. [10] used the trajectories from 100,000 cell phone users during six months to show a highly regulated human mobility pattern. Calabrese et al. [11] presented a method to extract mobility information from cell phone traces and established a multivariate regression model to predict human mobility. Song et al. [12, 13] proposed novel models to explore human mobility patterns.

As a main part of the public transportation system in cities, taxis provide a significant proportion of citizens' travel needs, due to their accessibility and flexibility. Furthermore, taking GPS-equipped taxis as probe vehicles, these mobile sensors provide us new tools to discover spatial-temporal patterns of people movement and even origin and destination distribution. Thus, compared to data sourced from cell phones, taxi locations data can reflect traveling characteristics more precisely, as passengers who use taxis have specific origins and destinations. Recently, lots of interesting works focus on human activity recognizing, hotspot discovering, urban planning, and transportation planning [17–28]. Liu et al. [17, 18] introduced a new method to explore intra-urban human mobility and land use variations based on taxi trajectory data from Shanghai. Castro et al. [19] proposed an overview of mechanisms for using taxi GPS data to analyze people's movements and activities, including three main categories: social dynamics, traffic dynamics, and operational dynamics.

Liang et al. [20] found the taxis' traveling displacements and elapsed time follow an exponential distribution instead of a power-law. In Refs. [21–23],

Veloso and Phithakkitnukoon used taxi data collected in Lisbon to study urban mobility, spatiotemporal variation of taxi services, relationships between pick-up and drop-off locations and drivers' behaviors. Zhu and Guo [24] proposed a hierarchical method to deal with the problem of how to extracts clusters from similar flows in taxi trips. Liu et al. [25] used a two-level hierarchical polycentric city structure to study spatial interaction perspective in Shanghai with large scale taxi data. Wu et al. [26] introduced a novel method to explore urban human mobility based on social media check-in data, in which they constructed transition probability to model travel demand distribution. Liu et al. [27] analyzed taxi drivers' spatial selection behavior, spatio-temporal operation behavior, route choice behavior, and operation tactics with taxi GPS traces. Pan et al. [28] discussed a new method by using taxi traces to classify the urban land-use features.

6.2 TRANSPORTATION DEMAND ANALYSIS AND ATTRACTIVENESS MODELING

6.2.1 Data Source

The taxi GPS data we used in this study are collected from about 1100 drivers in Harbin, a city located in the northeast of China. The data collection ran from July to December in 2012, the recording rate was 30 s, and total samples came to 2880 a day. Each data sample contains not only location information but also collecting time and status. Table 6.1 provides an overall description of taxi trajectory data. The "Time" indicates when the data was recorded, "Latitude" and "Longitude" provide location data of taxi vehicle, "Speed" is the velocity of the vehicle—the unit is kilometers per hour. "Orientation" represents driving direction, which is based on degrees from the North. "Status" represents whether the taxi is occupied by passengers—"0" shows the taxi is vacant and "1" means it is occupied.

6.2.2 Distribution Pattern of Demand

We classify taxi trips into two parts based on their status: (1) pick up and transport passengers from origin to destination; (2) roam on the road to find next passenger. The overall distribution of origins and destinations reflect the travel demand of citizens who use taxis as a transportation tool. In order to analyze the features of this demand, we firstly divide the main area of Harbin (longitude from 126.57 to 126.72 and latitude from 45.7 to 45.8) into 400 transportation districts. Each element contains an area of 0.015 (longitude) $\times$ 0.005 (latitude). Fig. 6.1 displays distributions of origins and destinations on weekday and weekend, the weekday we select is August 1 and the weekend is August 4 in 2012. The color bar in Fig. 6.1 represents the number of trips. The results show that most trip origins and destinations are located in the center of the city. Meanwhile, several zones in the south also attract large amounts of people. As we

TABLE 6.1 Data Sections of Taxi GPS Data in Harbin City

Taxi ID	Time	Latitude	Longitude	Speed	Orientation	Status
100,300,002	2012/8/1 6:59	45.738384	126.616920	35	109	0
100,300,002	2012/8/1 7:00	45.736588	126.614845	29	110	0
…	…	…	…	…	…	…
100,300,010	2012/8/1 11:08	45.757168	126.604280	40	8	1
100,300,010	2012/8/1 11:09	45.759000	126.605290	33	12	1
…	…	…	…	…	…	…

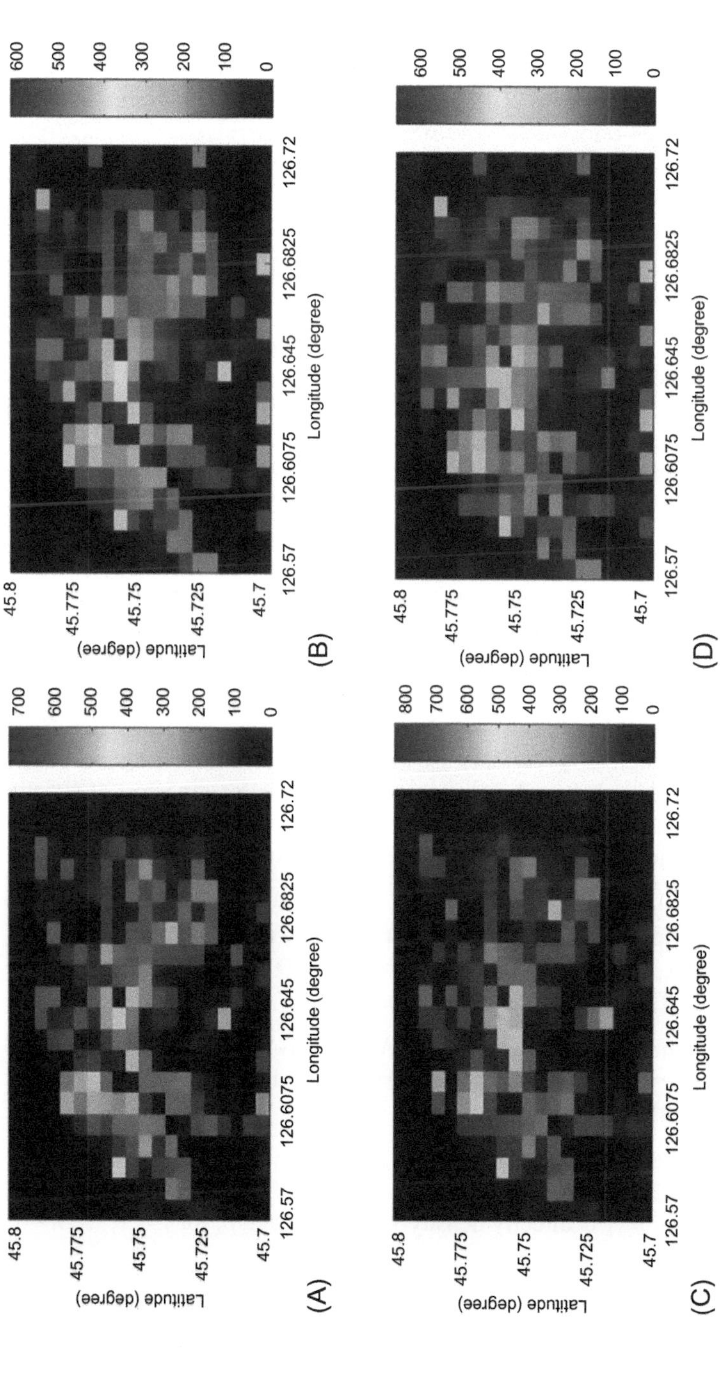

FIG. 6.1 Demand distribution of taxi trips: (A) origins on weekday, (B) destinations on weekday, (C) origins on weekend, and (D) destinations on weekend.

can see, the overall distributions of demand on weekday and weekend exhibit similar patterns except for some particular zones. Fig. 6.2 shows the hourly variety of origins and destinations on 2 days—x axis is the time horizon of 24 h and y axis is the number of trips. The total number of origins and destinations are 31,820 and 33,224 on weekday and weekend, respectively. The two distributions both express obvious peaks in the morning and evening, although the peak time on a weekday appears earlier than the weekend. This phenomenon is reasonable and consistent with urban travel patterns.

6.2.3 Clustering Based on DBSCAN

Although we obtain the distributions of origins and destinations, it is more important to understand which area in a city can attract more people and what the spatial distributions of these attracting locations are. In this section, we use density-based spatial clustering of applications with noise (DBSCAN) to cluster pick-up and drop-off locations and explore their spatial characteristics. The benefit of clustering includes the following two parts: (1) DBSCAN is a spatial density based method, and it can classify the locations in a cluster with high density, also the specific locations in the road network can be found for each cluster—these are the advantages compared to a grid based method; (2) DBSCAN can filter out the interfering noise. In China, passengers frequently call for vacant taxis in the middle link of a road; as these pick-up locations appear randomly, we should remove these locations for highly accurate clustering results. DBSCAN can realize this function through selecting proper parameters. The DBSCAN algorithm [29] is widely used in density-based clustering from large scale data for its simple calculation structure and low computing cost. It directly divides all point densities reachable from different points into clusters.

Before employing this algorithm, several definitions and terms should first be introduced.

Definition 1 (Eps-neighborhood of a point) The Eps-neighborhood of a point p, denoted by $N_{Eps}(p)$, is defined as $N_{Eps}(p) = \{q \in D \mid L_2(p, q) \le Eps\}$, where $L_2(p, q)$ is the Euclidean distance between p and q.

Definition 2 (directly density-reachable) A point p is directly density-reachable from a point q with respect to Eps and MinPts if and only if $p \in N_{Eps}(q)$ and $|N_{Eps}(q)| \ge MinPts$.

Definition 3 (density-reachable) A point p is density reachable from a point q with respect to Eps and MinPts if there is a chain of points $p_1, \ldots, p_n$, $p_1 = q$, $p_n = p$ such that p_{i+1} is directly density-reachable from p_i for $i = 1, \ldots, n-1$.

Definition 4 (density-connected) A point p is density connected to a point q with respect to Eps and MinPts if there is a point o such that both, p and q are density-reachable from o with respect to Eps and MinPts.

Definition 5 (cluster) Let D be a database of points. A cluster C with respect to Eps and MinPts is a nonempty subset of D satisfying the following conditions: (1) $\forall$ p, q: if $p \in C$ and q is density-reachable from p with respect to Eps and

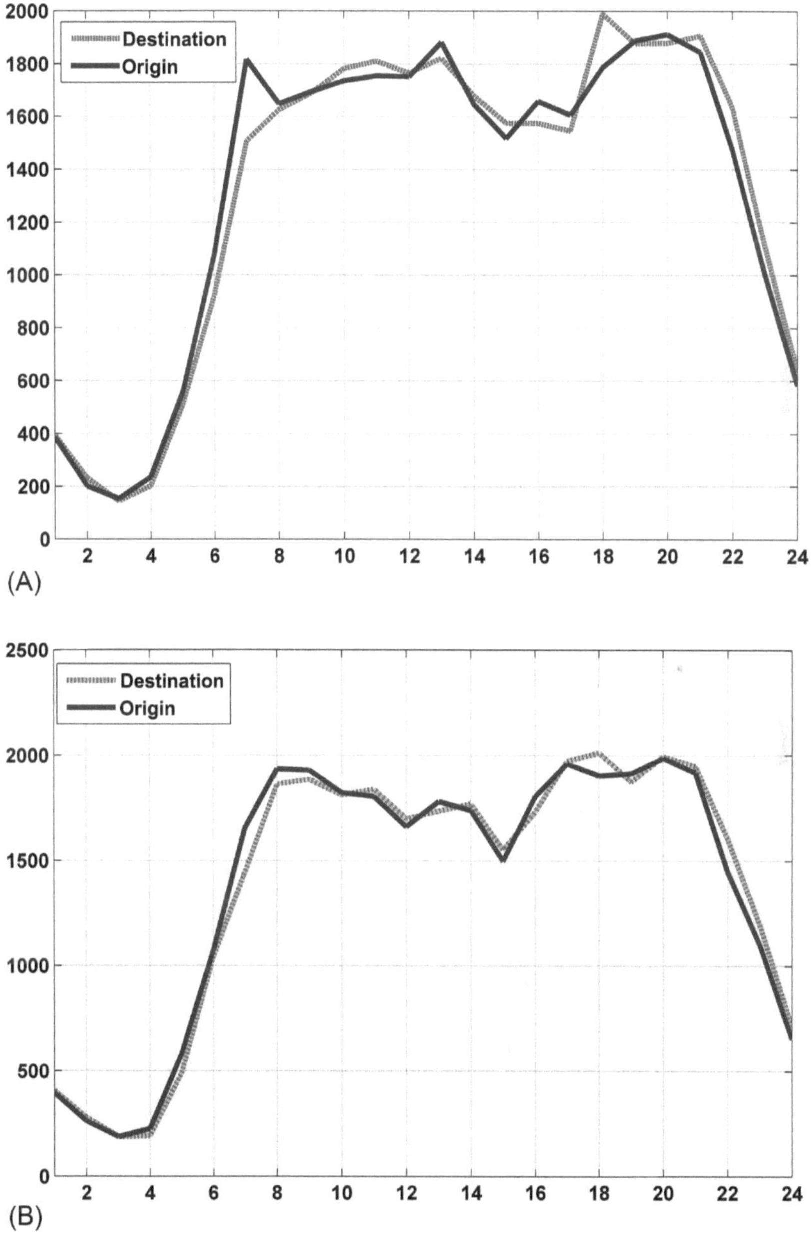

FIG. 6.2 Hourly taxi trip distribution for origins and destinations: (A) weekday and (B) weekend.

MinPts, then $q \in C$. (2) $\forall\ p, q \in C$: p is density-connected to q with respect to EPS and MinPts.

As we can see, two important parameters including the density threshold: MinPts and radius: Eps affect the clustering results in DBSCAN. Fig. 6.3A shows the changes of cluster number with different parameters based on the data

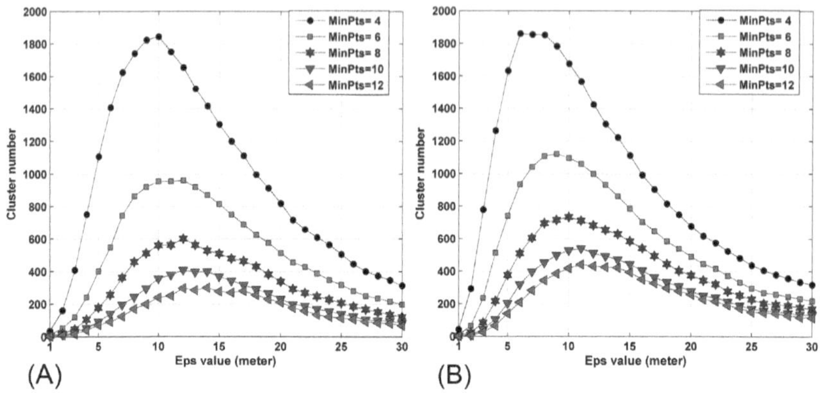

FIG. 6.3 Cluster numbers under different parameters: (A) pick-up locations and (B) drop-off locations.

collected from 1100 drivers during a week, from August 1 to 7. For a given MinPts, the cluster number rises gradually as Eps increases at the first section. Its value reaches to a peak and then decreases when Eps increases at the second section. The reason is that a small Eps value will cause the clusters to be separated and a large value means an initial lower density. When Eps is too large, the clusters will merge into a larger cluster. Another important parameter, MinPts, can also affect the clustering results for a given Eps. As the MinPts increases, the number of clusters decreases. A small MinPts value will result in more clusters being generated and a large value means the number of points clustered into the region with radius of Eps will increase. Furthermore, the curves of cluster number have similar distribution patterns when MinPts equal to 10–12, this means the cluster number becomes stable.

So, in this study, proper parameters are set as MinPts = 10 and Eps = 12, and 408 clusters are extracted from pick-up locations. Fig. 6.4A is the clustering results under the selected parameters, red points (gray in print versions) are the

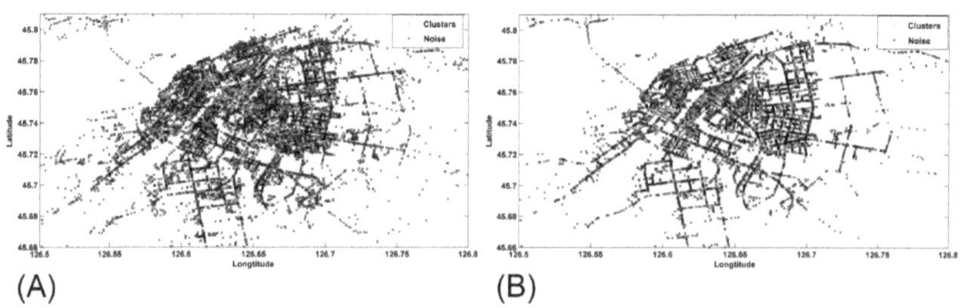

FIG. 6.4 Clustering results with defined parameters: (A) pick-up locations and (B) drop-off locations.

clusters, and black points represent the locations that cannot satisfy the clustering conditions in DBSCAN—they are generally treated as noise. By using a similar method, we can obtain the proper parameters (MinPts = 10, Eps = 11) and 538 clusters for drop-off locations shown in Fig. 6.3B. Fig. 6.4B shows the clustering results.

6.2.4 Attractiveness Model for Choosing Pick-Up Clusters

When taxi-drivers drop off a passenger, they try their best to find the next passenger around drop-off locations based on their driving experience. They consider candidate pick-up locations with two main factors: the number of passengers that may appear, based on their past experience, and the distance between drop-off and pick-up locations. Thus, the location with a shorter driving distance and more opportunity to find a customer will become a satisfying choice. In order to model this choice behavior, we should first estimate the centers of clusters and assume that search behavior happens between drop-off and pick-up centers.

We determine the cluster centers for each cluster under the condition of minimizing an objective function J. It is defined as:

$$J = \sum_{i=1}^{n} d(l_i, c) = \sum_{i=1}^{n} |l_i - c| \tag{6.1}$$

where $|\cdot|$ represents the general Euclidean distance, l is the location of the pick-up or drop-off, n is the number of locations in different clusters determined by DBSCAN, c means the cluster centers. First, we initialize the cluster center c. Then, iteratively modify the center to reduce the sum of the distances between each sample and the center. Finally, it terminates if one of the following conditions is satisfied: the value of objective function is below a certain tolerance; the difference of objective function between adjacent iterations is less than a presetting threshold; the iteration is complete. Here, we take a shopping center in Harbin for a case study. In Fig. 6.5, we extract 22 pick-up clusters and 9 drop-off clusters, the black points represent the calculated centers of pick-up cluster and the points marked with number represent the centers of drop-off cluster.We only show the choice distribution of drop-off centers 1 and 7 to pick-up centers.

A classical Huff model [30] is used to analyze drivers' choice behavior. The Huff model is a variant on the gravity and spatial interaction model and it measures the percentage of demand in each origin zone that will visit various destinations.

$$P_{ij} = \frac{T_{ij}}{\sum_{k=1}^{m} T_{ik}} = \frac{W_j^{\alpha} C_{ij}^{\beta}}{\sum_{k=1}^{m} W_k^{\alpha} C_{ik}^{\beta}} \tag{6.2}$$

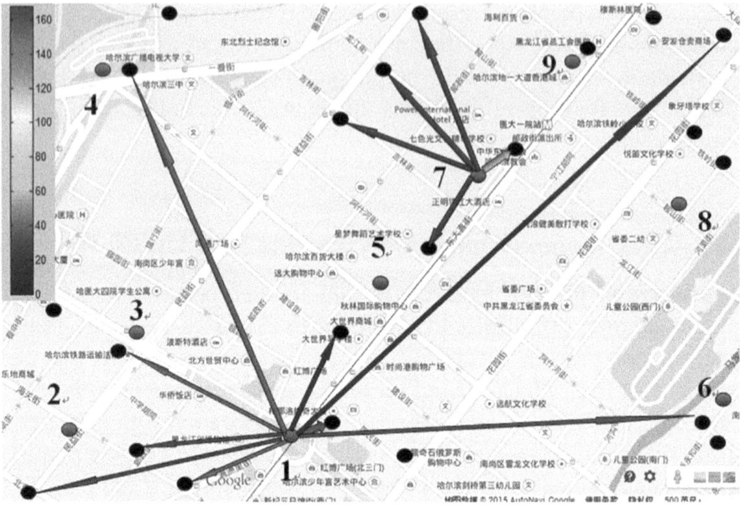

FIG. 6.5 A case study of a shopping center in Harbin city.

where P_{ij} is the probability of a taxi driver located at drop-off cluster center i choosing pick-up cluster center j; T_{ij} means the sum of trips from i to j; W_j is the attractiveness of pick-up cluster j (the number of pick-up locations in cluster is used to represent attractiveness); C_{ij} is the cost from i to j (the distance is used to estimate the cost); α and β are the sensitivity parameters; m is the number of pick-up cluster centers corresponding to drop-off center i. In Refs. [31, 32], four types of cost and attraction function combinations were compared to model attractiveness:

(1) $T_{ij} = \exp(\alpha \ln W_j - \beta \ln C_{ij}) = W_j^{\alpha} C_{ij}^{-\beta}$
(2) $T_{ij} = \exp(\alpha W_j - \beta C_{ij})$
(3) $T_{ij} = \exp(\alpha W_j - \beta \ln C_{ij}) = C_{ij}^{-\beta} \exp(\alpha W)$
(4) $T_{ij} = \exp(\alpha \ln W_j - \beta C_{ij}) = W_j^{\alpha} \exp(-\beta C_{ij})$

Here, we also compare the accuracy of four models. In order to calibrate the parameters, we construct an error function as follows:

$$E = \sum_{i=1}^{n} \sum_{j=1}^{m} \left(P_{ij}^{real} - P_{ij} \right)^2 \tag{6.3}$$

where P_{ij}^{real} means the observed choose probability; n is the number of drop-off cluster centers (in the case study $n = 9$); m is the number of pick-up cluster centers corresponding to drop-off center i. As E is a nonlinear objective function, the Levenberg-Marquardt (LM) method [33] is used to solve this nonlinear least square problem. The LM method is a widely used optimization algorithm in solving least square curve fitting and nonlinear programming problems.

TABLE 6.2 Parameters Estimation Results Based on LM Method

Models	α	β	Squared Sum of the Residual
1	1.0063	−0.2812	0.1163
2	0.0351	−0.0820	0.1727
3	0.0283	−0.3102	0.1471
4	0.9852	−0.0653	0.1318

Table 6.2 provides the calibration results based on the LM algorithm. The results show that the classic Huff model has the best fitting performance, with parameters $\alpha = 1.0063$ and $\beta = -0.2812$.

6.3 TRIPS DISTRIBUTION ANALYSIS

Taxi trips are a very important part of human beings' movements in urban areas. In this section, three parameters including travel distance, time, and average speed are used to explore taxi mobility. As we mentioned in Section 6.2, taxi drivers exhibit different driving behaviors depending on status: carrying passengers or vacant. Thus, the trips can be classified into two parts. Dataset for occupied taxi k at time period τ can be expressed as: $R^{\mathrm{o}} = (k, l^{\mathrm{o}}, \tau^{\mathrm{o}})$, in which $l^{\mathrm{o}} = (x^{\mathrm{o}}, y^{\mathrm{o}})$ includes longitude and latitude information. Similarly, the dataset of the nonoccupied taxi can be defined as $R^n = (k, l^n, \tau^n)$ and $l^n = (x^n, y^n)$. So, the travel distance is defined as:

$$d = \sum_{i=1}^{N-1} |l_{i+1} - l_i| \tag{6.4}$$

where N is the total number of data samples in a unique trip with status of occupied or nonoccupied, and $|\cdot|$ means the Euclidean distance of two adjacent locations. The travel time of a trip is defined as:

$$t = \tau_N - \tau_1 \tag{6.5}$$

where N is the total number of data samples in a unique trip with status of occupied or nonoccupied, τ_1 and τ_N are the start and end time of trip. The average speed of trip is defined as:

$$s = \frac{d}{t} \tag{6.6}$$

Finally, we extracts 31,823 and 31,828 trips on a weekday for occupied and nonoccupied status. On a weekend, 33,228 and 33,232 trips are extracted when the taxis is occupied and nonoccupied respectively.

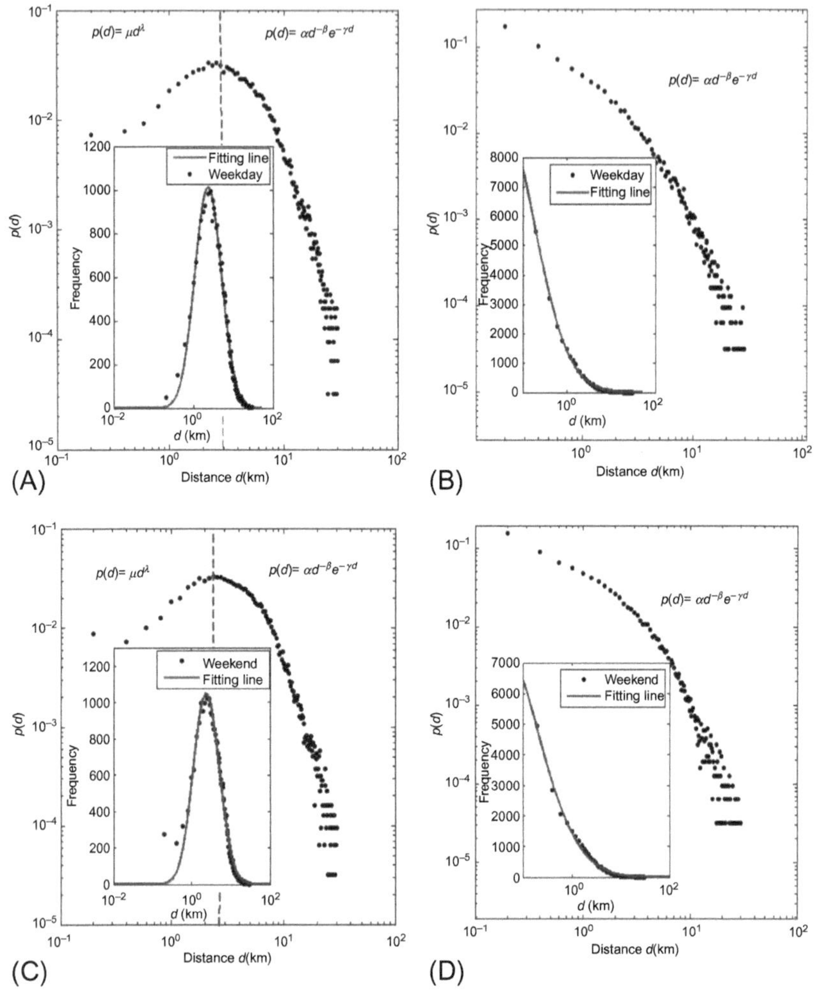

FIG. 6.6 Travel distance of trips. Weekday: (A) occupied trips and (B) nonoccupied trips. Weekend: (C) occupied trips and (D) nonoccupied trips.

6.3.1 Distance Distribution

To explore taxi mobility, we measure the frequency distribution with different travel distances and plot the probability $p(d)$ under double-logarithmic scale for the data collected on a weekday in Fig. 6.6A and B. By observing the figures, it is evident that the trips with occupied and nonoccupied status express different distributions. The $p(d)$ of occupied trips increases gradually and reaches the peak when d is about 3 km, and then descends as trip distance increases from 3 to 30 km. It indicates that most movements of passengers are limited in the urban area. We also can see that people seldom choose taxi to complete very short or very long distance travel. For the first part of the trips, the distribution of $p(d)$ can be approximated by a power-law function:

$$p(d) = \mu d^{\lambda} \tag{6.7}$$

For the second part of the trips, the distribution can be fitted to a power-law with exponential cut off (or truncated power-law):

$$p(d) = \alpha d^{-\beta} e^{-\gamma d} \tag{6.8}$$

All the parameters are shown in Table 6.3. The $p(d)$ of nonoccupied trips decreases as d increases from 0 to 30 km without increasing the trend. From the frequency distribution, we can see that smaller distances are most common across all trips. It indicates that in order to maximize profit, taxi drivers try their best to reduce "useless" travel distance. The decreasing trend means taxi drivers cruise through small distances around drop-off locations in order to quickly find new passengers. The distribution can be also fitted to the power-law with exponential cut off (or truncated power-law) in Eq. (6.8). The fitting parameters are provided in Table 6.3. For the trips extracted on a weekend, we obtain similar results, which can be seen in Fig. 6.6C and D, and Table 6.3.

We also fit the frequency distribution under logarithmic scale with log-normal function:

$$F(d) \propto e^{-\left(\frac{\log d - u_d}{\sigma_d}\right)^2} \tag{6.9}$$

The mean μ_d and standard deviation σ_d are optimized by Maximum Likelihood Estimate (MLE), the fitting lines and results are shown in Fig. 6.6 and Table 6.3.

6.3.2 Travel Time Distribution

Travel time is another important indicator to analyze human mobility. Geography determines the spatial information for travel distance, while travel time reflects travel accessibility and traffic conditions in the urban road network. We also plot the frequency and probability distribution of occupied and nonoccupied trips on weekday and weekend in Fig. 6.7.

The $p(t)$ of occupied trips increases gradually and reaches the peak when t is about 9 min, and then decreases as trip travel time increases from 9 to 100 min, see Fig. 6.7A. For the first part of the trips, the distribution of $p(t)$ can be approximated by a power-law function:

$$p(t) = \mu t^{\lambda} \tag{6.10}$$

For the second part of the trips, the distribution can be fitted to a power-law with exponential cut off:

TABLE 6.3 Fitting Parameters for Travel Distance Distribution

		p(d)					F(d)	
Status	Day	μ	λ	α	β	γ	μ_d	σ_d
Occupied	Weekday	0.023	0.635	0.132	0.497	0.221	0.377	0.478
	Weekend	0.025	0.573	0.191	0.732	0.206	0.381	0.466
Nonoccupied	Weekday	–	–	0.083	1.166	0.154	−1.566	1.128
	Weekend	–	–	0.088	1.095	0.161	−1.551	1.184

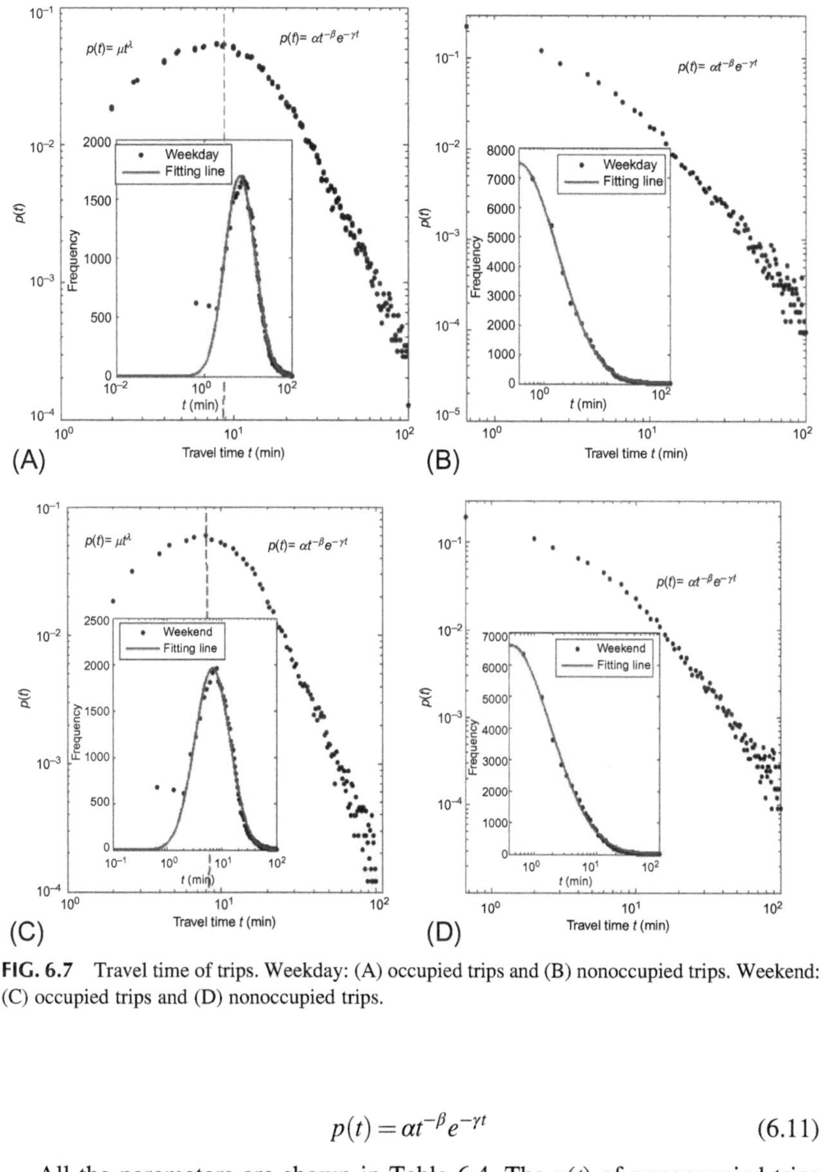

FIG. 6.7 Travel time of trips. Weekday: (A) occupied trips and (B) nonoccupied trips. Weekend: (C) occupied trips and (D) nonoccupied trips.

$$p(t) = \alpha t^{-\beta} e^{-\gamma t} \tag{6.11}$$

All the parameters are shown in Table 6.4. The $p(t)$ of nonoccupied trips decreases as t increases from 0 to 100 min, see Fig. 6.7B. The distribution can be also fitted by a power-law with exponential cut off in Eq. (6.11). The fitting parameters are provided in Table 6.4. For the trips extracted on weekend, the results and fitting parameters can be seen in the Fig. 6.7C and D, and Table 6.4. The frequency distribution under logarithmic scale can be fitted to log-normal function:

TABLE 6.4 Fitting Parameters for Travel Time Distribution

		$p(t)$					$F(t)$	
Status	Day	μ	λ	α	β	γ	μ_t	σ_t
Occupied	Weekday	0.017	0.528	3.921	1.438	0.019	0.831	0.483
	Weekend	0.015	0.689	7.354	1.667	0.018	0.828	0.453
Nonoccupied	Weekday	–	–	0.455	1.032	0.021	−0.407	0.874
	Weekend	–	–	0.501	1.157	0.019	−0.379	0.918

$$F(t) \propto e^{-\left(\frac{\log t - u_t}{\sigma_t}\right)^2} \tag{6.12}$$

The fitting lines and results are shown in Fig. 6.7 and Table 6.4.

6.3.3 Average Speed Distribution

We display the frequency and probability distribution of average speed on weekday and weekend in Fig. 6.8. The probability distributions of nonoccupied trips on weekday and weekend have obviously different patterns compared with

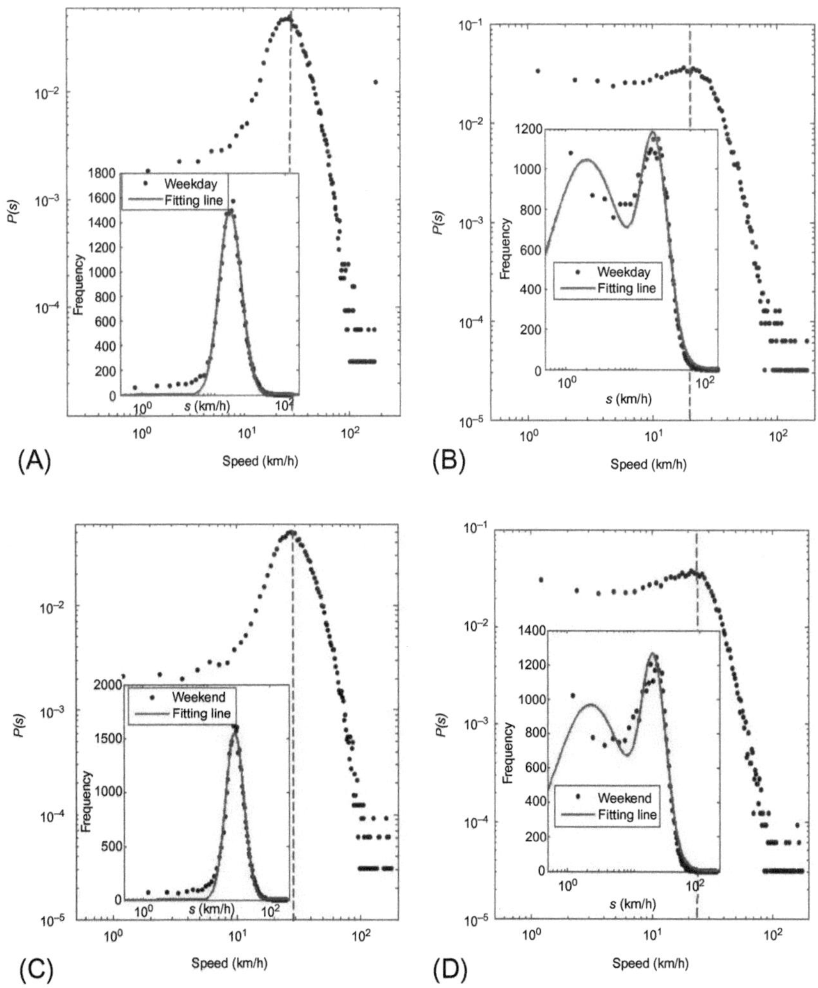

FIG. 6.8 Average speed of trips. Weekday: (A) occupied trips and (B) nonoccupied trips. Weekend: (C) occupied trips and (D) nonoccupied trips.

travel distance and time. The average speed of occupied trips is lower than that of nonoccupied trips, however, the proportion of low speed (<20 km/h) in non-occupied trips is higher than that in occupied trips. This reflects the different driving behavior in the two statuses. When the drivers deliver a passenger to a destination, they will quickly move to the possible pick-up locations based on their experience. Then they will slow down and cruise on the road near the attractive locations to carefully search for new passengers. In Fig. 6.8, using power-law or truncated power-law function to fit the distributions on double-logarithmic scale cannot provide satisfying fitting results. So, we only fit the frequency distribution based on log-normal function, for occupied trips:

$$F(s) \propto e^{-\left(\frac{\log s - u_s^1}{\sigma_s^1}\right)^2} \tag{6.13}$$

For nonoccupied trips, a combined function is used as following:

$$F(s) \propto e^{-\left(\frac{\log s - u_s^1}{\sigma_s^1}\right)^2} + e^{-\left(\frac{\log s - u_s^2}{\sigma_s^2}\right)^2} \tag{6.14}$$

The fitting parameters are shown in Table 6.5.

Unlike the results reported in Refs. [6, 8, 10, 16] that the power-law or truncated power-law distribution can be found from cell phone communication data, the occupied taxi trip distance and travel time distributions display combined patterns. The reason can be explained from following two aspects: (1) the size of urban area and structure of network limiting the travel space of taxi trips. Very long trips ($d > 15$ km, $t > 50$ min) are a small proportion of all the trips; (2) the economic condition is another factor. Taxi fares will increase as travel distance and time accumulate. Too short or too long trips are not economical. Generally, walking and cycling are the favorite transportation methods for short trips, and most people will choose subway or bus for long trips. In summary, as an important part of urban human movement, taxi trip distributions show unique characteristics, which not only reveal human mobility in the urban area but also provide constructive suggestions in transportation planning.

TABLE 6.5 Fitting Parameters for Average Speed Distribution

		F(s)			
Status	**Day**	μ_t^1	σ_t^1	μ_t^2	σ_t^2
Occupied	Weekday	1.226	0.221	–	–
	Weekend	1.439	0.204	–	–
Nonoccupied	Weekday	1.277	0.274	0.301	0.745
	Weekend	1.302	0.260	0.349	0.766

6.4 TRAFFIC DISTRIBUTION BASED ON ENTROPY-MAXIMIZING MODEL

Traffic distribution models can reflect the patterns of transport flow among origins and destinations. In this study, we use an entropy-maximizing model [34, 35] to estimate taxi traffic distribution in Harbin. Assuming g_i is the traffic generation probability of zone i, a_j is the traffic attraction probability of zone j, the distribution probability t_{ij} from i to j can be defined as:

$$\begin{aligned} g_i &= G_i/X, \quad \sum_i g_i = 1 \\ a_j &= A_j/X, \quad \sum_j a_j = 1 \\ t_{ij} &= X_{ij}/G_i, \quad \sum_j t_{ij} = 1 \end{aligned} \tag{6.15}$$

where, G_i is the traffic flow generated from zone i, A_j represents the traffic flow attracted to zone j, X_{ij} means the traffic flow distribute in OD pair (i, j), and X is the total number of trips. Furthermore, we can use observed data to calculate prior probability q_{ij} in Gravity model:

$$q_{ij} = \alpha g_i a_j d_{ij}^{-\gamma} \tag{6.16}$$

where d_{ij} means travel deterrence (measured by distance travel from i to j), α and γ are the fitting parameters, which can be calibrated using least square method. According to entropy-maximizing theory, the objective function is expressed as:

$$\text{Max } L = -\sum_i \sum_j g_i t_{ij} \ln(t_{ij}) - \gamma \sum_i \sum_j g_i t_{ij} \ln(d_{ij}) \tag{6.17}$$

$$\text{S.T.} \begin{cases} \sum_j t_{ij} = 1 \\ \sum_i g_i t_{ij} = a_j \end{cases}$$

The above programming problem can be solved, and we then obtain the t_{ij} as follows:

$$\begin{cases} t_{ij} = e^{-1} k_i m_j d_{ij}^{-\gamma} \\ \sum_j t_{ij} = 1 \\ \sum_i g_i t_{ij} = a_j \end{cases} \tag{6.18}$$

where $k_i = \exp.(\mu_i/g_i)$, $m_j = \exp.(\lambda_j)$, μ_i and λ_j are the Lagrange coefficient related to subject conditions, they should be determined by iterative calculation to satisfy the converge condition. Finally, the traffic distribution between OD pair (i, j) is estimated by the following equation:

$$X_{ij} = t_{ij} X_i \tag{6.19}$$

The main calculation process of the entropy-maximizing model is summarized as follows:

Step 1: estimate the value of γ in Eq. (6.16) and set initial values of μ_i and λ_j.

Step 2: update the values of μ_i based on the following equation:

$$\mu_i = -g_i \ln\left[\sum_j \exp\left(-1 + \lambda_j\right) d_{ij}^{-\gamma}\right] \tag{6.20}$$

Step 3: update the values of λ_j using μ_i calculated in the above step:

$$\lambda_j = \ln a_j - \ln\left[\sum_i f_i \exp\left(-1 + \mu_i / f_i\right) d_{ij}^{-\gamma}\right] \tag{6.21}$$

Step 4: judge whether the terminating condition is met, if the absolute difference of μ_i and λ_j in two adjacent iterations are both smaller than the presetting threshold ε, then calculate t_{ij} and X_{ij} in Eqs. (6.18), (6.19). If not, go to the next step.

Step 5: use the newest values of μ_i and λ_j to replace old ones, and repeat from Step2 to Step4 until the converge condition can be satisfied.

In the model application, we only use 12 zones in the central area of Harbin city. The reason is the large amount of elements in OD matrix generated from all zones (shown in Fig. 6.1) equal to zero. This OD matrix includes 400×400 elements and is a sparse matrix. People seldom choose taxi to complete long trip travel, and trip volumes between two zones is heavily influenced by their spatial distance. Thus, we can hardly obtain satisfying estimation results from a traffic distribution model by using this large, sparse matrix as observed samples. Furthermore, the purpose of application is to verify the estimation performance of the entropy-maximizing model, we consequently use these main 12 zones as our example (the high density zones shown in Fig. 6.1), and the OD matrix in this study includes 144 elements.

In the Gravity model, distance is the cost measurement and is calculated by spatial distance among centers of zones. The initial values of μ and λ are set to 1 and 0, respectively. The threshold ε is set to 0.0001. The calibrated parameters are shown in Table 6.6, in which the AME means absolute mean errors between estimated and observed values. Fig. 6.9A displays the estimation results of the calibrated entropy-maximizing model. The 144 elements in OD matrix as observed values, and the estimation error in Fig. 6.9B is the difference between estimated distribution probability and actual values. As we can see, the errors fluctuate above and below the zero line. The mean, standard deviation, minimum, and maximum values of errors are also shown in Table 6.6.

TABLE 6.6 Calibrated Parameters in Entropy-Maximizing Model

	Number of Zones	Max Iteration Steps	γ	Mean	Std. Deviation	Minimum	Maximum	AME
Values	12	5	−0.8836	0.0076	0.0661	−0.2665	0.2258	0.0407

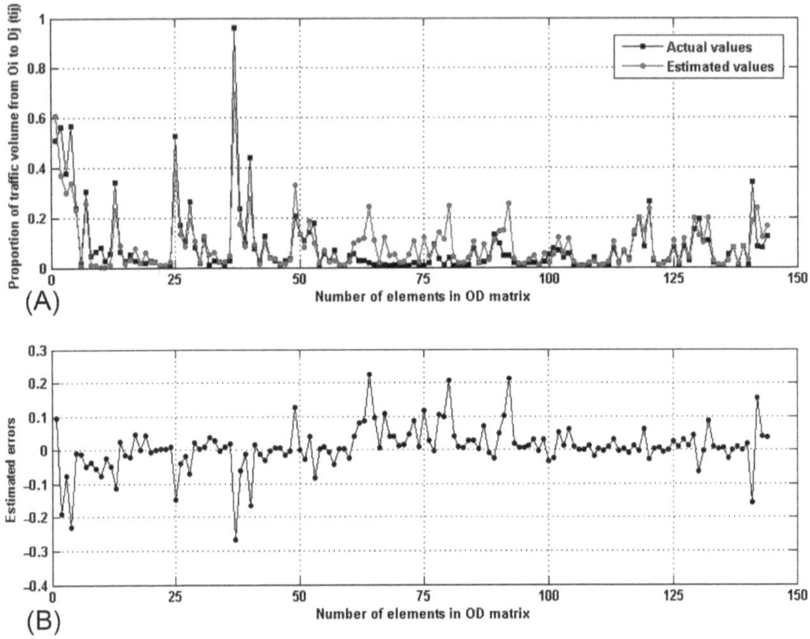

FIG. 6.9 Estimation results of traffic distribution using entropy-maximizing method: (A) comparison between estimated and observed values and (B) estimation errors.

6.5 NETWORK CONSTRUCTION AND DYNAMIC CHARACTERISTICS

In order to further explore the dynamic characteristics of the taxi trajectory data, we construct two travel networks based on the data recorded for occupied trips and vacant trips. These networks are referred to as the Occupied Trips based Travel Network (OTTN) and the Vacant Trips based Travel Network (VTTN), respectively. Before the network construction, a grid is overlaid on the main area of Harbin city (longitude from 126.55 to 126.73 and latitude from 45.68 to 45.8) and divided into 600 (30×20) cells. Each cell has a total land area of 600×600 m. For each travel network, considering each cell as a node, we can connect two nodes with an edge if there is a trip between them. The edge is directional, and the weights of edges can be defined as the number of trips between two nodes. Thus, these two networks can be viewed as directed and weighted graphs. In the following part of this section, we calculate some statistical properties of the network to explore taxi travel mobility.

6.5.1 Degree and Strength Distribution

The degree and the degree distribution are commonly used measurements to analyze complex networks [35–37]. The degree of a node is the number of edges incident with it, and the degree distribution $p(k)$ is defined as the probability that a node has degree k. The strength of node i is defined as the sum of the weights of all edges connected to it. Similarly, the strength distribution $p(s)$ is defined as the probability that a given node has strength s. Fig. 6.10 shows the cumulative probability distribution of the degree and the strength for the OTTN and VTTN. In the figure, the horizontal axis represents the degree and the strength of nodes, and the vertical axis represents the cumulative probability of degree and strength. In the figure, we can see the similarity between the distributions of degree and strength, both of which have obvious tails. A truncated power-law distribution, $p(x) \sim x^{-\alpha} f(\beta x)$, can further be fitted to each distribution.

We find that there are only a few nodes with large degree ($k > 500$) or high strength ($s > 10^4$), and most nodes have relative degree or strength values in the middle range of represented values; this means most cells in Harbin have close contact with surrounding cells. Since cells in the transportation grid can be treated as nodes in the travel network, degree and strength values can be used to evaluate the importance of cells (i.e., nodes) in terms of intensity (concentration of trips starting/ending in a given cell) and extent (portion of the network covered by taxi service). This also means most vacant trips in the VTTN end in cells nearby to those from which they originated. This distributional feature of vacant trips explains the fact that taxi drivers search for customers in areas of the grid near them (i.e., in adjacent cells) in order to save time. The distribution of occupied trips reflects the fact that the destinations of passengers are quite broad (home, shopping center, work place etc.)

6.5.2 Degree vs Strength Distribution

As the travel network is directed and weighted, we can further classify the strength into two categories: in-strength and out-strength. In-strength is defined as the sum of the weights of all incoming edges (i.e., the edges which consider the given node as a destination), and out-strength defined as the sum of the weights of all outgoing edges (i.e., the edges which consider the given node as an origin). The two strengths are denoted as:

$$s_i^{\text{in}} = \sum_{j=1}^{n} a_{ji} \cdot w_{ji} \tag{6.22}$$

$$s_i^{\text{out}} = \sum_{j=1}^{n} a_{ij} \cdot w_{ij} \tag{6.23}$$

where n is the number of nodes connected via edges to node i, w_{ij} is the weight of the edge that starts from node i and connects to node j, and a represents the adjacency matrix. Similarly, we can define the in-degree and out-degree of the network. In the roadway network of Harbin, the strength and degree can be used to evaluate the importance of nodes with respect to intensity and extent. Fig. 6.11

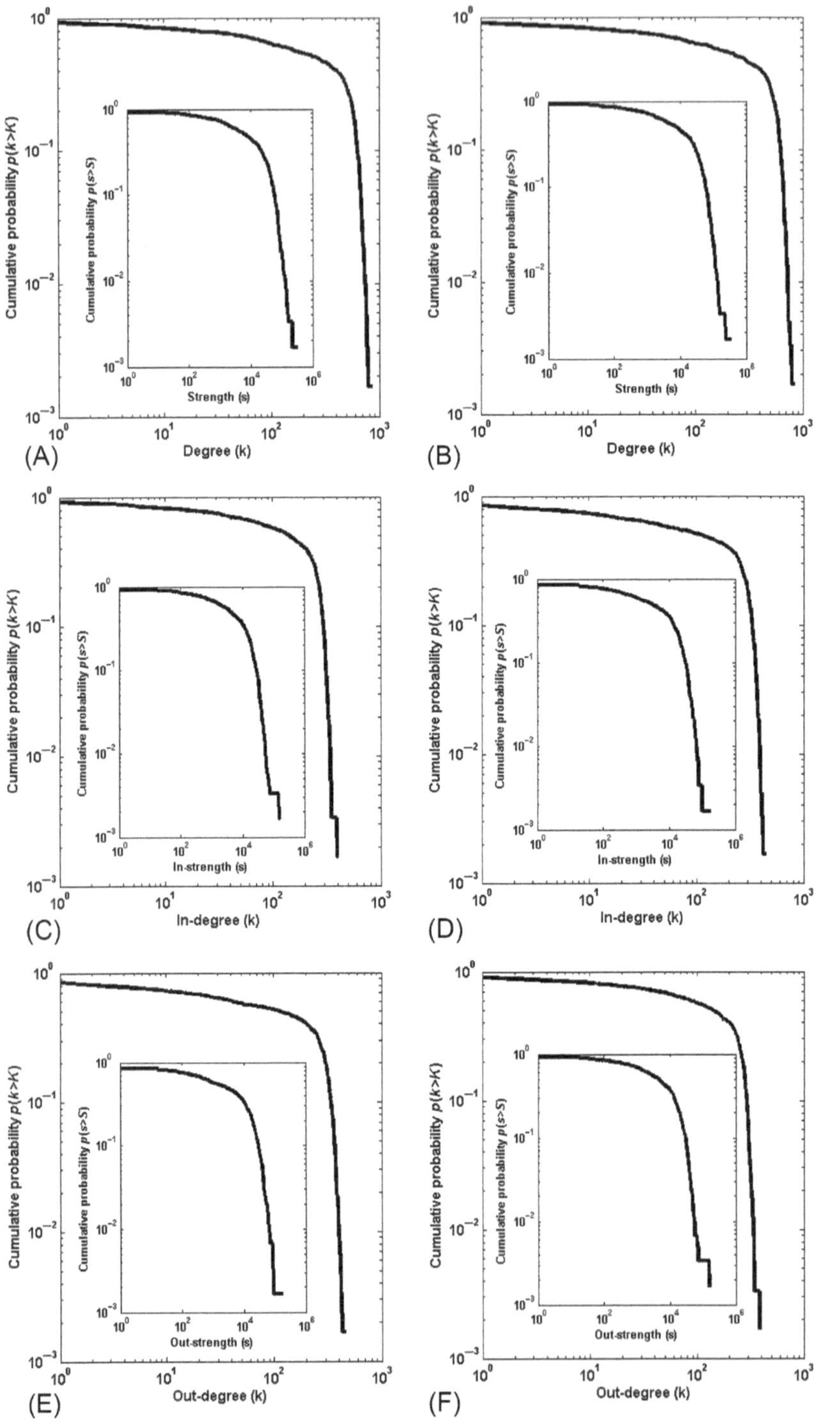

FIG. 6.10 Cumulative probability distribution of degree and strength: (A) degree and strength of occupied trips, (B) degree and strength of vacant trips, (C) in-degree and in-strength of occupied trips, (D) in-degree and in-strength of vacant trips, (E) out-degree and out-strength of occupied trips, and (F) out-degree and out-strength of vacant trips.

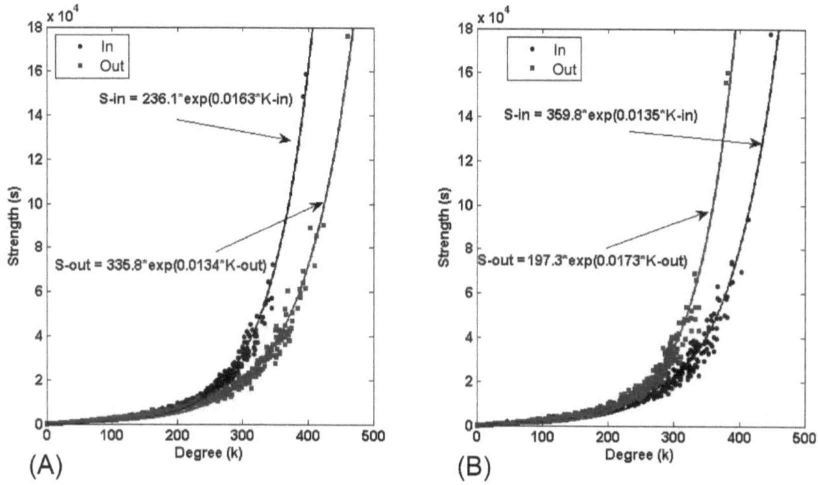

FIG. 6.11 Degree-strength correlation: (A) occupied trips and (B) vacant trips.

shows the degree-strength correlation for the two aforementioned networks (the OTTN and the VTTN). As can be seen, the plots of in-degree vs in-strength and the out-degree vs out-strength can both be characterized by exponential functions, as seen in Fig. 6.11.

The relationship between degree and strength is used to evaluate which nodes or cells are hotspots. The positive correlation between degree and strength represents that nodes with wide communications occupy an important component in the networks. Another interesting find is that the degree-strength relation for occupied trips has a different pattern than that for vacant trips. The in-degree of OTTN is lower than out-degree, while the in-degree of VTTN is higher than out-degree. In the occupied trips, taxi drivers transport customers from the original grid to other grids according to their destinations. Thus, for a single cell, its out-degree is generally higher than its in-degree. Similarly, for vacant trips, taxi drivers frequently drive into important cells in order to have a higher probability of picking up customers. So, for an individual cell, the in-degree is generally higher than the out-degree in the VTTN.

6.5.3 $k_i^{out}k_j^{in}$ vs w_{ij} Correlation

Fig. 6.12 a shows a plot of edge weight w_{ij} vs start-end degree $k_i^{out}k_j^{in}$, and from the figure, it can be observed that both the OTTN and VTTN exhibit similar patterns. Overall, it can be seen that the quantity $k_i^{out}k_j^{in}$ has some degree of positive correlation with edge weight. Further, the relationship between edge weight and start-end degree appears to be composed of two parts. When the value of $k_i^{out}k_j^{in}$ is small ($<10^5$), the weight w_{ij} increases slowly with increases in $k_i^{out}k_j^{in}$. When the value of $k_i^{out}k_j^{in}$ is large ($>10^5$), the rate of increase in

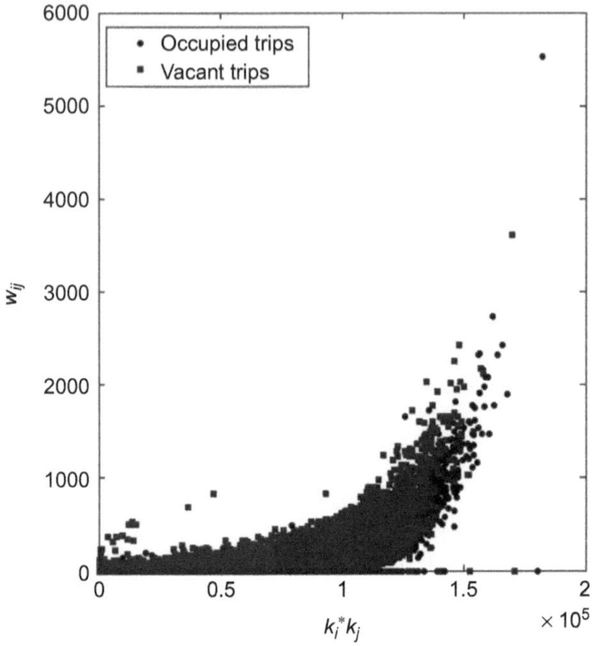

FIG. 6.12 Correlation between $k_i^{out}k_j^{in}$ and w_{ij}.

weight w_{ij} with increases in $k_i^{out}k_j^{in}$ becomes larger. This means that the nodes with high degree are likely to be connected to other nodes with high degree via high-weight edges. That is to say, in Harbin, an important area or cell has a higher tendency to be connected with other important areas or cells. This is a reasonable conclusion, since in occupied trips, the origins and destinations of customers are generally located in more important areas, and in vacant trips, taxi drivers also tend to find customers in hotspot areas.

6.5.4 Betweenness vs Strength and Clustering Coefficient

In a network, some nodes serve as connections between two communities, and removing these nodes will cause disconnection between the communities. In complex network theory, this characteristic is described by the indicator of betweenness, and the betweenness of node i is defined as:

$$B_i = \sum_{j \neq l \neq i} \frac{N_{jl}(i)}{N_{jl}} \tag{6.24}$$

where $N_{jl}(i)$ denotes the number of shortest paths from node j to node l passing through i and N_{jl} represents the total number of shortest paths from node j to node l. A node with high betweenness indicates that there are relatively short

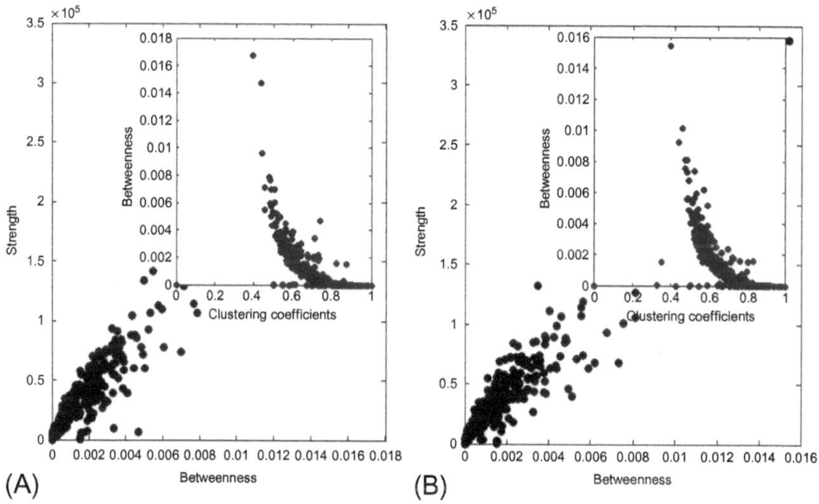

FIG. 6.13 Correlation between strength, clustering coefficients and betweenness: (A) occupied trips and (B) vacant trips.

paths between the subject node and others, and it also reflects the node's effect and influence in the network. Thus, in the taxi travel network, betweenness indicates the importance of locations, based on high density of customers. Fig. 6.13 shows the positive correlation between strength and betweenness, which indicates hotspots are well connected with other hotspots.

Furthermore, it can be observed in Fig. 6.13 that betweenness and the clustering coefficient appear to be negatively correlated. The clustering coefficient is the ratio of the number of existent connections between a given node's neighbors to the maximum number of possible connections between them, and it is considered a quantitative measure of the local properties of a complex network. If the clustering coefficient of some nodes equals one, it indicates that the neighbors of these nodes have the maximum possible number of connections between them. If the coefficient equals zero, it indicates that these nodes are isolated and have no connections with other nodes.

When plotting betweenness vs clustering coefficient as shown in Fig. 6.13, we can see that the nodes with large clustering coefficients tend to have small betweenness for both the OTTN and the VTTN. One can view nodes with large betweenness values as bridges connecting two clusters. Thus, compared to the nodes in clusters, the nodes between clusters usually have larger betweenness and smaller clustering coefficient values. In terms of transportation system, these nodes or cells play a key role in facilitating traffic flow within the network. If these nodes or cells are subject to abnormalities such as traffic congestion or accidents, the traffic flowing through some important regions in the network will definitely be impacted and likely delayed. Thus, these import nodes or cells should be taken into account during transportation planning efforts as failure to maintain connections between them can disrupt flow and cause breakdown at other significant cells in the network.

6.5.5 Network Construction and Structure Entropy

Summary statistics of both the occupied and vacant trip networks are shown in Table 6.7. From brief observation of the table, we can see that there are no obvious statistical differences between the occupied and vacant trips, and the OTTN and the VTTN appear to have a similar network structure, in which there are a large amount of nodes sharing connections by a low percentage of the total number of edges.

Furthermore, we use the quantity of entropy to evaluate the structural feature of the networks. Entropy plays a central role to information theory and can be used to measure uncertainty [38]. Accordingly, we define the network structure entropy (NSE) as follows. Let I_i be the importance of node i:

$$I_i = {}^{k_i}\Big/\sum_{i=1}^{n} k_i \tag{6.25}$$

where n is the total number of the nodes in network and k_i represents the degree of node i. Define E to be the NSE for the network:

$$E = -\sum_{i=1}^{n} I_i \cdot \ln I_i \tag{6.26}$$

As we can see, for a complete regular network, when the NSE reaches its maximum, we have

TABLE 6.7 Statistical Result of Two Travel Network

Network	Statistical Indicators	Degree	Edge Weight	Clustering Coefficients	Betweenness
Occupied trips	Mean	293.7221	17.1892	0.6515	0.0012
	Standard deviation	249.5639	57.3033	0.2475	0.0017
Vacant trips	Mean	287.1233	17.3967	0.6316	0.0010
	Standard deviation	245.1724	59.1724	0.2505	0.0016

$$E_{\max} = -\sum_{i=1}^{n} \frac{1}{n} \ln\frac{1}{n} = \ln n \tag{6.27}$$

On the other hand, we assume that a network has one core node with degree $n-1$, and all other nodes have degree 1. That is to say, $I(1)=1/2$, $I(i)=1/2(n-1)$ $(i>1)$, and the NSE reaches its minimum as follows:

$$E_{\min} = -\sum_{i=2}^{n} \frac{1}{2(n-1)} \ln\frac{1}{2(n-1)} - \frac{1}{2} \ln\frac{1}{2} = \frac{\ln 4(n-1)}{2} \tag{6.28}$$

In order to remove the impact of network size, the normalized NSE is defined as:

$$E_N = \frac{E - E_{\min}}{E_{\max} - E_{\min}} = \frac{-2\sum_{i=1}^{n} I_i \cdot \ln I_i - \ln 4(n-1)}{2\ln n - \ln 4(n-1)} \tag{6.29}$$

Obviously, $0 \leq \text{EN} \leq 1$.

In general, a "small world" network is a kind of network with a small average shortest path length and high clustering coefficients. The average shortest path length of a network is defined as:

$$Length = \frac{1}{N(N-1)} \sum_{i,j \in N, i \neq j} d_{ij} \tag{6.30}$$

where N is the total number of nodes in the network and d_{ij} represents the shortest path length between nodes i and j. Based on the aforementioned definitions, we find that the average shortest path length is 4.7611 and the clustering coefficient is 0.6515 for the OTTN. For the VTTN, the average shortest path length is 4.7591 and the clustering coefficient is 0.6316. These results indicate that neither of the two networks considered in this study exhibit the "small world" property. This is because the low clustering coefficients show that there are many small clusters in each network, and a lot of nodes cannot be directly connected, thus resulting in an increase in average distance within the network.

In addition, since the shortest average distance of the network is calculated based upon all reachable node pairs, and this network contains some isolated nodes, the network studied herein is different from other social networks. Therefore, it is necessary to consider using another index to evaluate characteristics of the network structure. The normalized NSE, En, is a widely used indicator to analyze network structure. In terms of the definition of NSE, the value of EN is small for a network with good connectivity. When the network is split into several random local networks, or lots of nodes have no links connecting them with the other nodes, the value of EN will increase. We find that the values of EN for two travel networks considered in this study are both high. This means the network is split into several random local networks, or lots of nodes have no

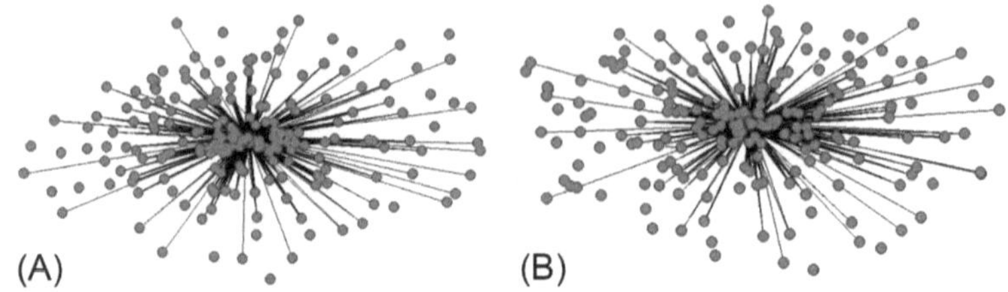

FIG. 6.14 Network structure of OTTN and VTTN: (A) occupied (EN=0.8259) and (B) vacant (EN= 0.8032).

links with the other nodes. In summary, according to the analysis of the aforementioned three quantitative indices, we infer that the travel networks do not have the "small world" characteristic (Fig. 6.14).

A selection of statistical properties of networks was calculated to study urban travel mobility, such as degree, strength, betweenness, clustering coefficient, edge weight, and NSE. From these features, we can understand the nature of hotspots in Harbin and more thoroughly explore travel mobility based upon taxi trajectories. Additionally, travel behavior for occupied and vacant trips can be studied using analysis techniques from the field of complex networks. In the next sections, we will identify traffic zones based on community detection algorithms and model the distribution of taxi trips.

6.6 SPATIAL-TEMPORAL PROPERTIES OF URBAN TRAVEL

6.6.1 Traffic Zone Identification

Based on the previous discussion, we adopt community detection algorithms to identify traffic zones from a network-based perceptive. Identification of these zones will further help to uncover spatial interactions of human movements in an urban system. Currently, there are a variety of algorithms used for community detection within complex networks. In this study, we apply and compare two widely used methods of community detection in order to identify traffic zones: the Louvain Method [36] and the visualization of similarities (VOS) method [37]. The aim of the Louvain method is to optimize modularity as the algorithm progresses. VOS is a new mapping technique that can serve as an alternative to the well-known technique of multidimensional scaling. The idea of VOS is to minimize a weighted sum of the squared distances between all pairs of items. The squared distance between a pair of items is weighted by the similarity between the items [37]. VOS Quality is used to evaluate the similarity between the two methods used for community detection.

In the main area of Harbin, there are five significant regions: Qunli, Xiangfang, Nangang, Daoli, and Daowai; these regions are shown in Fig. 6.15. In order to estimate the similarity between the administrative division of regions and the recognition results from the two algorithms, we adopt two indices

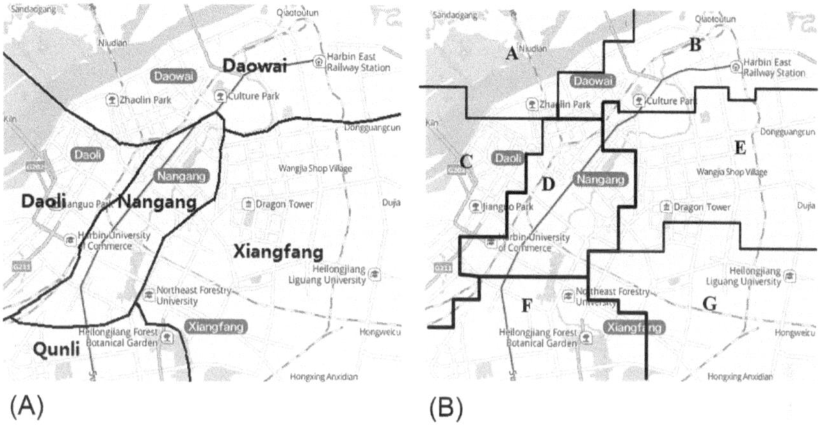

FIG. 6.15 Regional partition based on Louvain method in main area of Harbin city: (A) administrative divisions and (B) recognized by identification algorithms.

[38–40]: the Rand index (RI) and the Fowkles-Mallows index (FMI). The Rand index [39] takes values between zero and one, with zero indicating that the two data clusters do not agree on any pair of points and one indicating that the data clusters are exactly the same. FMI [40] is an external evaluation method that is used to determine the similarity between two clusters. Furthermore, as FMI considers both the matching and mismatching parts of two clusters, values of FMI are generally lower than those of the RI. Table 6.8 shows the detection results of communities in the OTTN and the VTTN. VOS provides finer recognition results and detects more communities than the Louvain method. In Fig. 6.8, the identification results of regions in the network constructed from occupied trips using the Louvain method only are shown. We can see that the identification results are quite similar with the actual administrative divisions: communities A and B compose Daowai, C and D represent Daoli and Nangang, E and G form Xiangfang, and F represents Qunli. The values of RI and FMI in

TABLE 6.8 Community Detection Results

Methods	Travel Networks	No. of Communities	Evaluation Index	Similarity Index	
				Rand	*FM*
Louvain	Occupied	7	$Q=0.0415$	0.8126	0.5873
	Vacant	7	$Q=0.0686$	0.8065	0.5620
VOS	Occupied	15	Quality = 0.3008	0.7931	0.5519
	Vacant	10	Quality = 0.3042	0.7754	0.5412

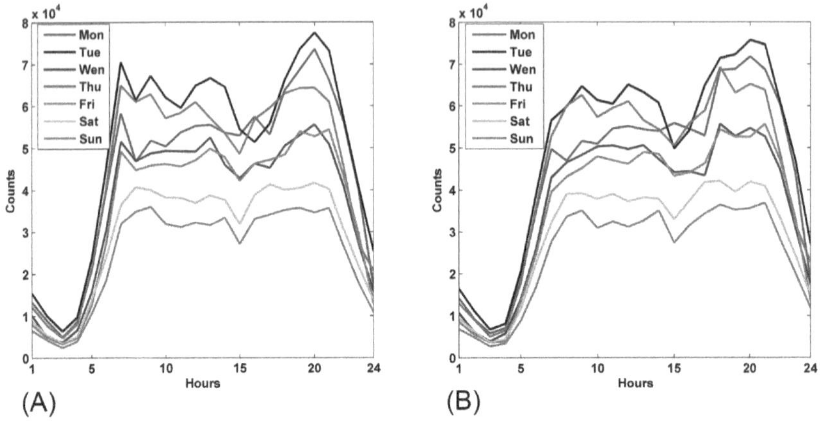

FIG. 6.16 Hourly variation of trip numbers in a week: (A) occupied trips and (B) vacant trips.

Table 6.8 also demonstrate the high coincidence between traffic zones identified from taxi GPS trajectories and the actual administrative divisions.

6.6.2 Travel Pattern Analysis

In order to analyze general travel patterns for each day of the week, we quantify the average volume of occupied and vacant trips over the half year period during which taxi trajectory data were collected in Harbin. Fig. 6.9 shows the hourly variations in the numbers of occupied and vacant trips for each day of the week. In the figure, the x axis indicates the hour within a given day and the y axis shows the number of trips. The number of trips on weekdays is obviously higher than that on weekends. Furthermore, for both weekdays and weekends, there are two peaks that can be observed: (1) the morning peak starting at around 8 am and 7 am, respectively, for occupied and vacant trips; and (2) the evening peak starting at around 8 pm for both types of trips. The volume of both types of trips reaches its minimum value at approximately 3 pm on each day. This distribution of trips by time is consistent with urban travel patterns, namely the morning and evening commute between residential areas and workplaces (Fig. 6.16).

In terms of examining the spatial construction of a city, Louail et al. [41, 42] applied an urban dilatation index: the average weighted distance between individual cell phone users, to analyze the organization of the city. Considering each occupied taxi trip as an individual traveler's trip, we adopt this method to explore spatial characters of Harbin. Before implementing this approach, several definitions should first be introduced.

The average distance between individuals is evaluated by the Venables index [41, 43], which can be defined as:

$$V=\sum_{i\neq j} s_i s_j d_{ij} \tag{6.31}$$

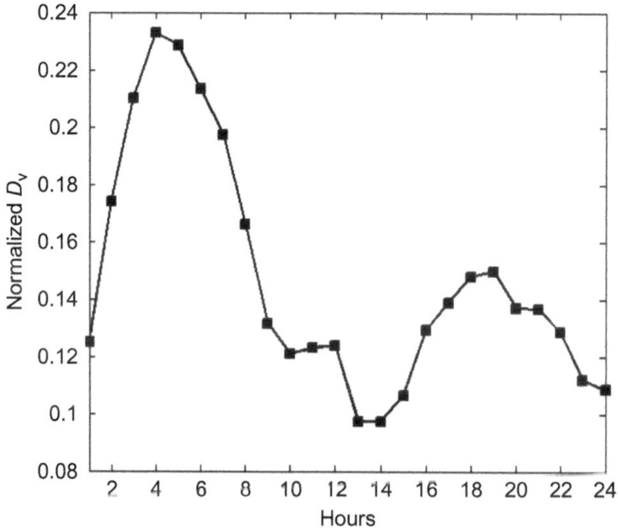

FIG. 6.17 Hourly variation of normalized D_V on weekdays.

where s_i represents the percentage of individuals in cell i at time t, and d_{ij} is the distance between cells i and j. We can imagine that if all taxis are staying in one cell or spatial unit, the value of V will reach its minimum. Furthermore, the distance V can be normalized by weights to obtain the weighted average distance, D_V, which is expressed as follows:

$$D_V(t) = \frac{\sum_{i<j} s_i(t)s_j(t)d_{ij}}{\sum_{i<j} s_i(t)s_j(t)} \tag{6.32}$$

In order to compare values of D_V for different cities, the value can further be normalized by the square root of the area of the city: $D_V/\sqrt{A}$. Fig. 6.10 shows the distribution of normalized average weighted distance over the 24 h in a day. Here, occupied trips, which can reflect the distribution of origins and destinations of citizens, are used to calculate D_V. There are two significant variations in the plot of normalized D_V vs time corresponding with the morning and evening peaks (i.e., the commutes from and to the suburbs). According to the conclusion in Ref. [41], the high values of D_V and its obvious fluctuation over the course of the day reveal that Harbin is a typical monocentric city (Fig. 6.17).

6.6.3 Hotspot Analysis

Hotspots are defined as locations that attract a lot of citizens within a given location; they can further indicate points of interest and travel patterns in urban areas, as well as other useful information. Identifying hotspots is valuable to transportation planning, especially transit planning. The process of hotspot identification includes the following two steps: (1) use the origins and destinations extracted from taxi GPS records to calculate local density, which is the number of trips starting or ending in cell i at time t: $\rho_{ori}(i,t)$ and $\rho_{des}(i,t)$—the locations or areas in the destination dataset can be used to analyze attractiveness from other locations or areas; similarly, the locations in the origin dataset can be used to explore the ability of an area to generate trips to other locations; (2) determine proper thresholds and identify hotspots—a cell can be estimated as a hotspot if its local density is larger than the preset threshold δ.

According to the work in Ref. [41], we apply two thresholds in this study, namely δ_{min} and δ_{max}. The threshold δ_{min} is simply defined as the average local density, and δ_{max} is determined from a Lorenz curve. A Lorenz curve is a graphical representation of the distribution of income or wealth commonly used in the field of economics. In order to plot a Lorenz curve, we first sort the local densities of all cells in ascending order: $\rho_{ori}(1,t) < \rho_{ori}(2,t) < \cdots < \rho_{ori}(n,t)$ or $\rho_{des}(1,t) < \rho_{des}(2,t) < \cdots < \rho_{des}(n,t)$, where n is the number of cells. In the Lorenz curve, the x axis is the proportion of cells: $P = i/n$, and the y axis is the cumulative proportion of density, which is expressed as follows:

$$L(i,t) = \frac{\sum_{j=1}^{i} \rho(j,t)}{\sum_{j=1}^{n} \rho(j,t)} \tag{6.33}$$

If all densities are equal and hence sorted at the same position, the Lorenz curve is plotted as a diagonal line. Usually, however, curvature can be observed in the plot, which indicates inequality in terms of density across cells. The stronger the curvature in a Lorenz curve, the higher the degree of inequality. Fig. 6.18

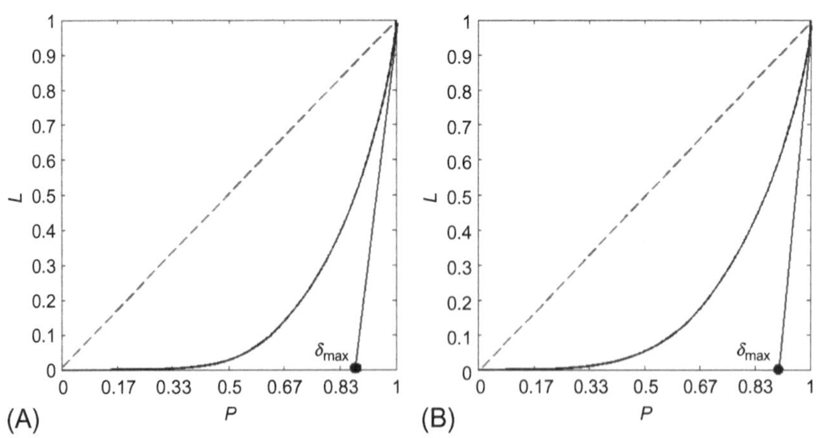

FIG. 6.18 Threshold selection in Lorenz curves: (A) origins and (B) destinations.

shows the Lorenz curves developed based on the origin and destination datasets. The red line is the diagonal line with slope equal to 1 from [0, 0] to [1, 1], and the black curve is the Lorenz curve. Here, we follow the procedure outlined in Ref. [41] to estimate δ_{max}: the intersection between yhr tangent of $L(P = 1)$ (the straight line at right corner in the figure) and the x axis. More detail on this procedure for interested readers can be found in Ref. [41].

Based on the aforementioned method, we finally determine the thresholds for origin locations: $\delta_{min} = 30{,}500$, $\delta_{max} = 78{,}900$ and destination locations: $\delta_{min} = 29{,}864$, $\delta_{max} = 88{,}000$. Fig. 6.12 displays the density distribution in 600 cells and can be used to identify hotspots based on the two sets of thresholds for origins and destinations, respectively. The number of hotspots meeting the criteria of the upper thresholds is obvious less than those meeting the criteria of the lower ones. Furthermore, from the observation of the distribution of hotspots, some cells in a city can not only attract large amounts of people, but also generate large numbers of trips. Such cells are important locations in Harbin and should be considered as central to transportation and transit planning efforts (Fig. 6.19).

6.7 CONCLUSIONS

In this chapter, we characterize human mobility from urban taxi trips. We divide a city area into 400 unique transportation districts and find out the distribution patterns of origins and destinations on a weekday and weekend. The DBSCAN algorithm is used to cluster pick-up and drop-off locations, and two key parameters (MinPts and Eps) in the algorithm are optimized. Accordingly, four spatial interaction models are calibrated and compared based on trajectories in the shopping center of Harbin city to study the pick-up location searching behavior; we find that the classical Huff model has the best modeling accuracy. Furthermore, we find that distribution of taxi trips in occupied status include two patterns: an ascending part and a descending part. The distribution in the ascending part can be well fitted to a power law function, and the curve in the descending part follows a truncated power law function. As to distribution of trips in non-occupied status, there only exists a monotonically pattern, which can be fitted to a truncated power law function. Furthermore, in the case studies, we optimized the parameters in an entropy-maximizing model based on actual OD matrix and evaluated its estimation accuracy for traffic distribution. Additionally, we apply and compare two widely used community detection algorithms, the Louvain Method and the VOS method, to identify traffic zones in the main area of Harbin. The communities identified by the algorithms are similar with the actual administrative divisions of Harbin. Finally, the spatial structure of Harbin is analyzed by using the dilatation index, and hotspots in the urban area are estimated based on the locations of trip origins and destinations.

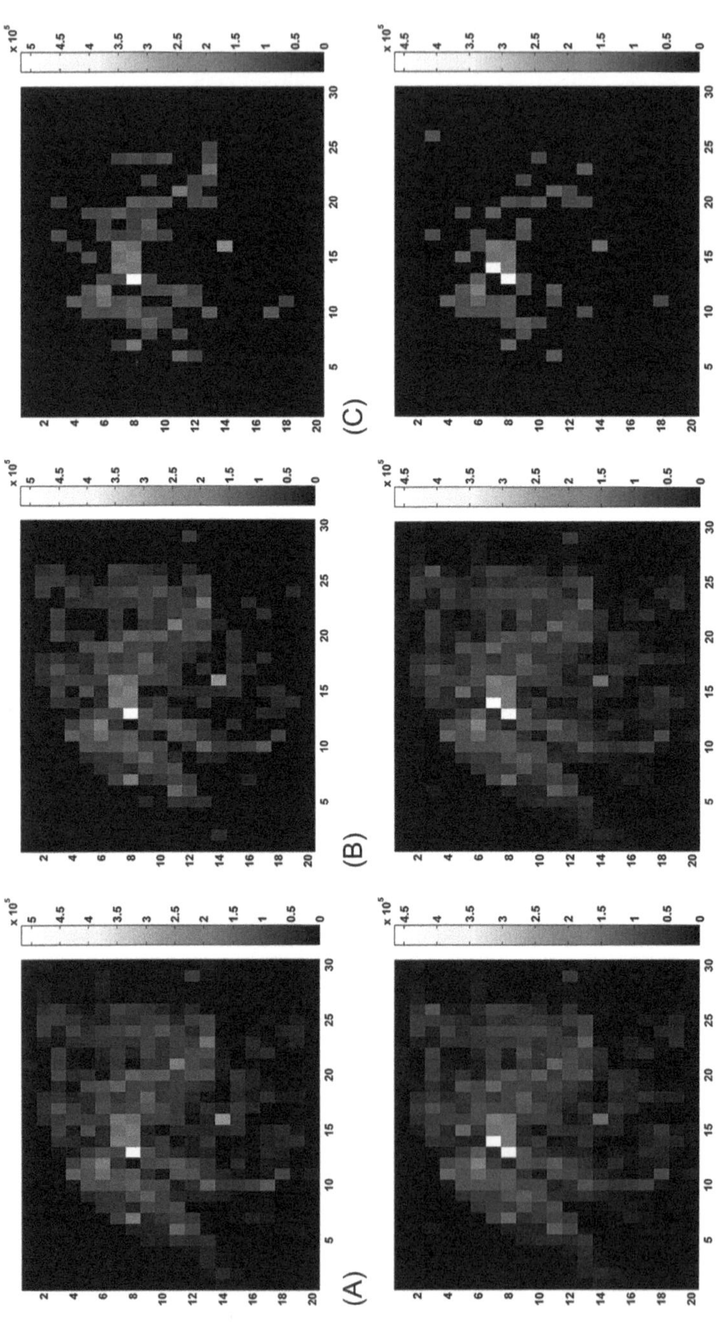

FIG. 6.19 Identification of hotspots with two different criteria: (A) density of origins, (B) hotspots of origins with δ_{min}, (C) hotspots of origins with δ_{max}, (D) density of destinations, (E) hotspots of destinations with δ_{min}, and (F) hotspots of destinations with δ_{max}.

REFERENCES

[1] M. Kim, D. Kotz, S. Kim, in: Extracting a mobility model from real user traces, Proceedings of the 25th IEEE International Conference on Computer Communications, 2006, pp. 1–13.

[2] B. Jiang, J. Yin, S. Zhao, Characterizing the human mobility pattern in a large street network, Phys. Rev. E 80 (2009) 021136.

[3] Y. Zheng, Q. Li, Y. Chen, X. Xie, W.-Y. Ma, in: Understanding mobility based on GPS data, Proceedings of the 10th International Conference on Ubiquitous Computing, Seoul, Korea, 2008, pp. 312–321.

[4] A. Bazzani, B. Giorgini, S. Rambaldi, R. Gallotti, L. Giovannini, Statistical laws in urban mobility from microscopic GPS data in the area of Florence, J. Stat. Mech: Theory Exp. 2010 (2010) P05001.

[5] Q. Li, T. Zhang, H. Wang, Z. Zeng, Dynamic accessibility mapping using floating car data: a network-constrained density estimation approach, J. Transp. Geogr. 19 (3) (2011) 379–393.

[6] B.C. Csáji, A. Browetc, V.A. Traag c, J.-C. Delvenn, E. Huensc, P.V. Doorenc, Z. Smoredae, V.D. Blondel, Exploring the mobility of mobile phone users, Phys. A 392 (2013) 1459–1473.

[7] J.B. Sun, J. Yuan, Y. Wang, H.B. Si, X.M. Shan, Exploring space-time structure of human mobility in urban space, Phys. A 390 (2011) 929–942.

[8] C. Kang, X. Ma, D. Tong, Y. Liu, Intra-urban human mobility patterns: an urban morphology perspective, Phys. A 391 (2012) 1702–1717.

[9] C.A. Hidalgo, C. Rodriguez-Sickert, The dynamics of a mobile phone network, Phys. A 387 (2008) 3017–3024.

[10] M.C. González, C. Hidalgo, A.-L. Barabási, Understanding individual human mobility patterns, Nature 453 (7196) (2008) 79–82.

[11] F. Calabrese, M. Diao, G.D. Lorenzo, J. Ferreira, C. Ratti, Understanding individual mobility patterns from urban sensing data: a mobile phone trace example, Transp. Res. C 26 (2013) 301–313.

[12] C. Song, Z. Qu, N. Blumm, A.-L. Barabási, Limits of predictability in human mobility, Science 327 (2010) 1018–1021.

[13] C. Song, T. Koren, P. Wang, A.-L. Barabási, Modeling the scaling properties of human mobility, Nat. Phys. 6 (2010) 818–823.

[14] R. Ahas, A. Aasa, S. Silm, M. Tiru, Daily rhythms of suburban commuter's movements in the Tallinn metropolitan area: case study with mobile positioning data, Transp. Res. C 18 (1) (2010) 45–54.

[15] A. Sevtsuk, C. Ratti, Does urban mobility have a daily routine? Learning from the aggregate data of mobile networks, J. Urban Tech. 17 (1) (2010) 41–60.

[16] D. Brockmann, L. Hufnagel, T. Geisel, The scaling laws of human travel, Nature 439 (2006) 462–465.

[17] Y. Liu, C. Kang, S. Gao, Y. Xiao, Y. Tian, Understanding characteristics of intra-urban trips using taxi trajectory data, J. Geogr. Syst. 14 (4) (2012) 463–483.

[18] Y. Liu, F. Wang, Y. Xiao, S. Gao, Urban land uses and traffic 'source-sink areas': evidence from GPS-enabled taxi data in Shanghai, Landsc. Urban Plan. 106 (1) (2012) 73–87.

[19] P.S. Castro, D. Zhang, C. Chen, S. Li, G. Pan, From taxi GPS traces to social and community dynamics: a survey, ACM Comput. Surv. 46 (2) (2013) 17.

[20] X. Liang, X. Zheng, W. Lu, T. Zhu, K. Xu, The scaling of human mobility by taxis is exponential, Phys. A 391 (2012) 2135–2144.

[21] M. Veloso, S. Phithakkitnukoon, C. Bento, in: Sensing urban mobility with taxi flow, Proceedings of the 3rd ACM SIGSPATIAL International Workshop on Location-Based Social Networks, ACM, New York, NY, 2011, pp. 41–44.

[22] S. Phithakkitnukoon, M. Veloso, C. Bento, A. Biderman, C. Ratti, in: Taxi-aware map: identifying and predicting vacant taxis in the city, First International Joint Conference on Ambient Intelligence (AmI'10), 2010, pp. 86–95. November 10–12, Malaga, Spain.
[23] M. Veloso, S. Phithakkitnukoon, C. Bento, Urban mobility study using taxi traces, in: Proceedings of the 2011 International Workshop on Trajectory Data Mining and Analysis, ACM, New York, NY, 2011, pp. 23–30.
[24] X. Zhu, D. Guo, Mapping large spatial flow data with hierarchical clustering, Trans. GIS 18 (3) (2014) 421–435.
[25] X. Liu, L. Gong, Y. Gong, Y. Liu, Revealing travel patterns and city structure with taxi trip data, J. Transp. Geogr. 43 (2015) 78–90.
[26] L. Wu, Y. Zhi, Z. Sui, Y. Liu, Intra-Urban human mobility and activity transition: evidence from social media check-in data, PLoS One 9 (5) (2014) e97010.
[27] L. Liu, C. Andris, C. Ratti, Uncovering cabdrivers' behavior patterns from their digital traces, Comput. Environ. Urban. Syst. 34 (6) (2010) 541–548.
[28] G. Pan, G. Qi, Z. Wu, D. Zhang, S. Li, Land-use classification using taxi GPS traces, IEEE Trans. Intell. Transp. Syst. 14 (1) (2013) 113–123.
[29] M. Ester, H. Kriegel, J. Sander, X. Xu, in: A density-based algorithm for sparse representations, Proceedings of the 2nd International Conference on Knowledge Discovery and Data Mining, 1996, pp. 226–231.
[30] D.L. Huff, A probabilistic analysis of shopping center trade areas, Land Econ. 39 (1) (1963) 81–90.
[31] M.E. O'Kelly, Trade-area models and choice-based samples: methods, Environ. Plan. A 31 (4) (1999) 613–627.
[32] Y. Yue, H. Wang, B. Hu, Exploratory calibration of a spatial interaction model using taxi GPS trajectories, Comput. Environ. Urban. Syst. 36 (2) (2012) 140–153.
[33] J. Nocedal, S.J. Wright, Numerical Optimization, Springer, New York, NY, 2006, pp. 258–264.
[34] A. Wilson, Entropy in urban and regional modeling: retrospect and prospect, Geogr. Anal. 42 (4) (2010) 364–394.
[35] H. Yang, T. Sasaki, Y. Iida, Estimation of origin-destination matrices from link traffic counts on congested networks, Transp. Res. B Methodol. 26 (6) (1992) 417–434.
[36] V.D. Blondel, J.L. Guillaume, R. Lambiotte, et al., Fast unfolding of communities in large networks, J. Stat. Mech: Theory Exp. 10 (2008) P10008.
[37] N.J. van Eck, L. Waltman, R. Dekker, et al., A comparison of two techniques for bibliometric mapping: multidimensional scaling and VOS, J. Am. Soc. Inf. Sci. Technol. 61 (12) (2010) 2405–2416.
[38] C. Kang, S. Sobolevsky, Y. Liu, et al., in: Exploring human movements in Singapore: a comparative analysis based on mobile phone and taxicab usages, Proceedings of the 2nd ACM SIGKDD International Workshop on Urban Computing, ACM, 2013.
[39] W.M. Rand, Objective criteria for the evaluation of clustering methods, J. Am. Stat. Assoc. 66 (336) (1971) 846–850.
[40] E.B. Fowlkes, C.L. Mallows, A method for comparing two hierarchical clusterings, J. Am. Stat. Assoc. 78 (383) (1983) 553–569.
[41] T. Louail, M. Lenormand, O.G.C. Ros, et al., From mobile phone data to the spatial structure of cities, Sci. Rep. 4 (2014) 5276.
[42] T. Louail, M. Lenormand, M. Picornell, et al., Uncovering the spatial structure of mobility networks, Nat. Commun. 6 (2015) 6007.
[43] R.H.M. Pereira, V. Nadalin, L. Monasterio, et al., Urban centrality: a simple index, Geogr. Anal. 45 (1) (2013) 77–89.

Chapter 7

Public Transportation Big Data Mining and Analysis

Xiaolei Ma and Xi Chen

School of Transportation Science and Engineering, Beihang University, Beijing, People's Republic of China

Chapter Outline

7.1 INTRODUCTION

As city populations, number of motor vehicles, and residents' travel demands increase, urban road traffic flows are becoming ever more saturated with frequent urban traffic congestion. Urban transportation has become a problem that needs to be addressed urgently during the development process of many large- and medium-sized cities throughout the world. Based on practical experience accumulated worldwide with respect to urban transportation, prioritizing the

Data-Driven Solutions to Transportation Problems. https://doi.org/10.1016/B978-0-12-817026-7.00007-2

development of urban public transportation is an effective solution to the urban transportation problem. In recent years, the strategic position of public transportation in urban transportation has become increasingly critical, with an urgent need for constructing urban public transportation systems with high service levels and operating efficiencies to satisfy the development demand of urban transportation.

In recent years, with the emergence and continuous advancement of concepts such as intelligent transportation and smart cities, the urban transportation industry has established a relatively mature data acquisition and technology application foundation. With the advent of the big data era, massive, multi-source spatial and temporal data, which are represented by cell phone data, floating car data, and social networks, are continuously employed in urban planning practice.

Manual survey data have been the primary data source for the analysis of the demand-and-supply relationship of public transportation systems and their operation and planning. Comprehensive spatial and temporal public transportation ridership distribution data are difficult to obtain by conventional transport survey methods. Conventional survey methods often collect residents' travel information via questionnaires or queries. Residents' travel data acquired using these methods generally do not satisfy the quantity and quality requirements for sample selection. These data have a series of problems, such as small survey sample sizes, low accuracy, high cost, low temporal effectiveness, long survey cycles, and difficulties in later-stage data processing. In addition, these data do not reflect the dynamic variation characteristics of passengers' travel demand.

Big data, urban public transportation smart cards, and satellite positioning technology, as well as informationalized survey and statistical methods for ridership, provide convenience for passengers. Massive data are generated during the public transportation operation process. These data are highly continuous, have wide coverage, contain comprehensive information, and update dynamically and quickly; thus, they have high application value to public transportation operation and planning.

During the urban public transportation operation process, the public transportation information infrastructure accumulates massive static and dynamic data in addition to providing transport information services. In this chapter, case studies are provided using data from the Beijing public transportation system to introduce the application of public transportation big data to practical problems. First, the basic types of public transportation data are introduced briefly. The following data types are commonly employed.

(1) Global positioning system (GPS) data

GPS data refer to the vehicle movement information acquired in real time by the GPS sensors installed on public transportation vehicles, such as taxis, buses, and subways.

(2) Ridership data

Urban residents choose different means of public transportation for their trips. The operational data of these means of public transportation constitute ridership data. Bus and subway ridership data are acquired by recording the usage of public transportation smart cards. These ridership data have high analytical value and can be used to monitor population movements, evaluate the urban public transportation system, and study the behavior of passengers.

(3) Mobile terminal data

Cell phones and tablet computers (e.g., iPad) are the mainstream mobile communication tools in today's society. Users can search for various types of transport information in real time with their cell phones and tablet computers. This mode of enquiry provides various types of data. Many public transportation systems (including bus routes) have launched social networking services to gather residents' feedback in real time to make timely adjustments and improvements.

(4) Video surveillance data

Video surveillance technology has been applied extensively in transport management. Massive video data collected by video surveillance equipment each day record urban residents' travel zones, which form a virtual "map" of residents' trips within a physical city in digital space. Cameras are also installed by the boarding and alighting doors of buses, and are primarily employed to capture video images of passengers. Data acquisition personnel determine the number of boarding and alighting passengers by analyzing and processing continuous video images using software.

(5) Wireless-fidelity (Wi-Fi) data

Existing dynamic urban public transportation monitoring systems rely on a third-generation (3G) network, which had shortcomings such as slow data transmission and high communication costs. A Wi-Fi-based on-board data transmission scheduling strategy for public transportation vehicles can reduce the total energy consumption of vehicles and data transmission time. Wi-Fi-based wireless transmission of on-board data of a public transportation vehicle refers to the transmission of updated data that are required by the vehicle via the File Transfer Protocol (FTP) by an on-board embedded device that connects to the server via the local Wi-Fi network.

This chapter is organized as follows: Section 7.2 introduces a quality control method for public transportation big data via case studies using data from the Beijing public transportation system. Section 7.3 introduces the application of public transportation smart card data in the determination of travel behavior and places of work and residence of commuters. Section 7.4 introduces the application of public transportation smart card data to the prediction of bus arrival times. Section 7.5 presents a public transportation big data analysis and visualization system that was developed based on an open-source geographic information system (GIS) platform under the theoretical framework of *E*-Science, which is used to realize the application of big data in public transportation operation and planning.

7.2 PUBLIC TRANSPORTATION BIG DATA PREPROCESSING METHOD

The previous section briefly introduced types of data from the Beijing public transportation system. In the actual public transportation data mining process, researchers need to clean the original data after acquiring them. Data cleaning is the last step in the process of discovering and correcting identifiable errors in data files, including data consistency checking and handling of invalid and missing values. Because smart card and GPS data that are commonly involved in public transportation data mining are large in volume, they require special processing software for management, cleaning, and storage. This section primarily introduces cleaning and preprocessing methods for public transportation smart card and GPS data via case studies using data from the Beijing public transportation system.

7.2.1 Public Transportation Smart Card Data Cleaning

Mining and analysis of public transportation smart card data (Fig. 7.1) has two objectives: to assist the transit operators in making decisions about public transportation; and to provide a data basis for public transportation planning. Public transportation data preprocessing is a process of obtaining data that can be used for the subsequent data mining process by screening and correcting the data in the established database and statistical analysis tools. The following data-related problems should be considered when preprocessing public transportation smart card data [1, 2].

(1) Missing data

An occurrence of missing data will have an adverse impact on later-stage data processing. If key fields are missing, the analysis process may be impeded. Occurrences of missing public transportation smart card data are relatively uncommon. When certain data are missing, known data should be employed as a substitute for the missing data. If no data can be used as a substitute, inference may be made based on experience or known data.

(2) Data errors

Public transportation smart card data may contain some erroneous statistical data, e.g., time recorded as "24:30:00." To ensure that data analyses do not yield erroneous results, data validity should be examined and redundant data should be deleted. Classification and sorting methods should also be employed to discover the data that may cause noise in the analysis results.

	card ID	Type	Transaction date	transaction time	Route number	bus ID	Boarding stop	Exit stop	Driver ID	Conductor ID
1	1000751063284790	06	20150601	063611	24038	00047665	0	1	04003743	04005511
2	1000751021088624	08	20150601	101233	00849	00042305	23	10	10001507	10025646
3	1000751006774985	08	20150601	004336	01113	00008035	28	24	00001092	00001068
4	1000751084565249	08	20150601	002950	01113	00008035	24	12	00001092	00001068
5	1000751088078348	08	20150601	220031	01113	00008035	24	12	00001092	00001068
6	1000751069782551	08	20150601	003320	01113	00008035	24	11	00001092	00001068
7	1000751039790296	06	20150601	101839	00381	00023331	0	1	02000807	02010561
8	1000751039158930	06	20150601	102437	00381	00023331	0	1	02000807	02010561
9	1000751070149415	06	20150601	102408	00381	00023331	0	1	02000807	02010561
10	1000751001709850	06	20150601	120208	00381	00023331	0	1	02000807	02010561

FIG. 7.1 Example of public transportation smart card data.

(3) Redundant data

Redundant data contain repeated information. For example, repeatedly recorded data are the most simple and visual redundant data. The presence of redundant data increases data accuracy. When some data contain errors, information can be recovered using the redundant data. However, this redundant data will cause difficulties in later-stage data processing or erroneous analysis. For example, repeated smart card data recordings will cause an overestimation of ridership. Redundant data should be treated based on the specific objective of data analysis. In addition, space consumption due to data storage should also be considered.

(4) Data consistency

The source of public transportation system data involve multiple devices and departments. Data acquisition targets vary among different device manufacturers and departments, which generates a difference in data with the same meaning among different devices and departments. Data inconsistency may be an inconsistency in data accuracy, data units, or data storage formats, or an inconsistency in data definitions. Data unit and storage format problems can be addressed by standardization at the data preprocessing stage. However, data accuracy and definitions cannot be addressed after the data are collected. A fundamental solution to data accuracy and definition problems is to use a unified data specification to specify clearly data definitions, accuracy, and units.

(5) Stale data

As time progresses, some data that were acquired a long time ago may become stale and invalid. Stale data comprise a problem relative to the objective of data analysis. Stale data can be stored in a separate subdatabase.

(6) Removing useless fields

Datasheets contain numerous fields, some of which have no significance to the data analysis in this chapter. Therefore, these redundant fields can be removed to accelerate the data analysis process. The following fields have direct significance to the data analysis in this chapter: card identification number (card ID), transaction date, transaction time, public transportation smart card type, vehicle ID, route number, and record number. Thus, the fields amount received, card balance, city number, card issue number, monthly pass type, amount due, driver ID, and conductor ID can be deleted to simplify and accelerate the data analysis process.

7.2.2 GPS Data Cleaning

GPS data are primarily collected by on-board GPS systems. The GPS data format may be inconsistent with the actual data format that is required by the study. Therefore, the GPS data format needs to be converted prior to the data analysis process.

The GPS data of the Beijing public transportation system, as shown in Fig. 7.2, are selected as an example. These GPS data reflect some problems. Therefore, the following aspects should be considered when preprocessing these GPS data [3]:

(a) The route numbers obtained from parsing are incorrect. Correct route number and vehicle ID information should be determined based on the subscriber identity module (SIM) card information and the information in the vehicle information table.
(b) The up/down signs are incorrect. As a result, the traveling directions of the buses cannot be determined.
(c) The direction data should be multiplied by 10. For example, the direction that corresponds to 21 is 210 degrees.
(d) The coordinates should be translated by adding 0.006 to the longitude and adding 0.001 to the latitude to ensure the compatibility of the coordinate data with the GIS data of the bus routes.
(e) The number of information items recorded in each GPS file is limited. As a result, the GPS files are not completely compatible with the smart card data.
(f) The time recorded in each GPS file is Greenwich Mean Time (GMT). Therefore, 8 h should be added to the originally recorded time to convert it to Beijing Standard Time.

This section introduces preprocessing methods for public transportation smart card data and original GPS data. Processed data can be mined and analyzed. Starting in the following section, the application of public transportation big data in practical problems is introduced from the aspects of public transportation system planning, operation, management, and evaluation.

7.3 APPLICATION OF PUBLIC TRANSPORTATION DATA IN PLANNING

Based on the introductions in the previous two sections, public transportation smart card data contain rich travel information and represent the travel behavior of each passenger. Therefore, determining the spatial and temporal patterns of passengers and identifying their travel patterns can provide a basis for urban public transportation planning, public transportation stop layout optimization, and public transportation development, as well as data support for departments such as the public transportation group, traffic control department, and development and planning department in relevant ridership analysis, which can facilitate reasonable optimization of bus and rail transport. The previously mentioned analysis is important for identifying and analyzing the commuting behaviors of public transportation and the operational scheduling of public transportation for morning and evening peak hours. Therefore, this section explains how to mine and analyze the commuting behaviors of transit

	Up/down sign	time	Positioning state	latitude	longitude	speed	directon	route number
1	4	20150518080533	A	39.8660966666667	116.440315	0.1852	28	69300
2	4	20150518080537	A	39.8660766666667	116.440291666667	0.7408	30	69300
3	4	20150518080603	A	39.8659883333333	116.440183333333	0.926	0	69300
4	4	20150518080618	A	39.8659833333333	116.440161666667	0.5556	2	69300
5	4	20150518080621	A	39.8659883333333	116.440166666667	0.5556	2	69300
6	4	20150518080648	A	39.865985	116.440161666667	0.1852	3	69300
7	4	20150518080718	A	39.8659166666667	116.440108333333	0.3704	31	69300
8	4	20150518080722	A	39.8659316666667	116.44012	0.1852	30	69300
9	4	20150518080748	A	39.8660083333333	116.440178333333	0.1852	28	69300
10	4	20150518080818	A	39.866005	116.440163333333	0.1852	24	69300

FIG. 7.2 Example of original GPS data of the Beijing public transportation system.

passengers based on public transportation data via case studies using Beijing public transportation smart card data.

7.3.1 Extraction of the Commuting Characteristics of Public Transportation Passengers

The term "commute" refers to the travel of a person between his/her place of residence and place of work. Four aspects are considered when extracting commuting characteristics: the number of travel days and departure time from a temporal pattern perspective, and the commuter stops and commuter routes from a spatial pattern perspective. Methods for extracting commuting characteristics and calculating the commuting indices are as follows.

Step 1: Same departure times

Departure times are measured in half-hours. One day is divided into 48 periods and denoted by 0–47 (0 signifies that the departure time falls between 0:00 a.m. and 0:30 a.m.; 47 signifies that the departure time falls between 11:30 p.m. and 0:00 p.m.). The most frequent departure time for the home-to-work trip (T_h) and the most frequent departure time for the work-to-home trip (T_w) are extracted.

For example, for the holder of the card numbered 5389, $T_h = 12$ with a total of 9 occurrences of T_h, and $T_w = 18$ with a total of 10 occurrences of T_w. Thus, a total of $9 + 10 = 19$ occurrences of the most frequent departure time is observed.

Step 2: Number of travel days (N_{day})

N_{day} is the actual number of days when the commuter travels compared with the number of times that the card has been used.

For example, card ID 5389 has been used 50 times on 15 separate days in June. Therefore, $N_{day} = 15$.

Step 3: Same commuter stops

Stops are classified into two types: place-of-work stops and place-of-residence stops. The origin (O) and destination (D) of the first trip are considered the place of residence and place of work, respectively. The O and D of the last trip are considered to be the place of work and place of residence, respectively. The most frequent place-of-work stop (S_w) and the most frequent place-of-residence stop (S_h) are extracted.

For example, the information extracted from the usage of card ID 5389 indicates that the card has been used 25 times at the most frequent place-of-residence stop and 26 times at the most frequent place-of-work stop. Thus, the card has been used $25 + 26 = 51$ times at the most frequent stops.

Step 4: Same travel routes

For the same commuter routes, the most frequent routes of the first and last trips and the number of trips within a month are recorded.

Step 4.1: To address the problem in which the commuter takes different routes from the same O to the same D, the number of trips is supplemented based on the identified most frequent place of residence and most frequent place of work.
Step 4.2: The most frequent routes of the first and last trips are removed.
Step 4.3: The number of routes of the first trip with O as the most frequent place of residence and D as the most frequent place of work is determined. The number of routes of the last trip with O as the most frequent place of work and D as the most frequent place of residence is determined.
Step 4.4: The sum of the previously mentioned four types of routes is calculated.

For example, the information extracted from the usage of card ID 5389 indicates eight occurrences of the most frequent route for the home-to-work trip, six occurrences of the most frequent route for the work-to-home trip, two occurrences of a nonmost frequent commuter route for the home-to-work trip, and four occurrences of a nonmost frequent commuter route for the work-to-home trip. Thus, a total of $8 + 2 + 6 + 4 = 20$ occurrences of the four types of routes are observed. Table 7.1 lists an example of extraction of commuting indices.

The holder of card 32,664 is now taken as an example. This commuter often travels from his/her place of residence (stop 265) to his/her place of work (stop 723) by taking route 57,300 and then transferring to route 60,366 between 8:00 a.m. and 8:30 a.m. In the afternoon, this commuter travels from his/her place of work to his/her place of residence by taking route 60,366 and then transferring to route 00741.

7.3.2 Identification of Commuters and Estimation of Their Places of Work and Residence

Commuters are determined and identified using a rating method based on the four commuting indices (parameters) extracted in the previous section. The determination of commuters consists of two parts. The first part is to determine the baseline score for the rating process based on the iterative self-organizing data analysis technique (ISODATA) algorithm [4], which is employed to assist the determination of the commuter baseline. The baseline score is also used to classify commuters into different groups. For example, based on their scores, commuters are classified into absolute commuters, average commuters, and noncommuters. The second part is to rate each commuter based on the technique for order of preference by similarity to ideal solution (TOPSIS) algorithm [5] and calculate relevant scores based on the four indices for commuting characteristics, from which the travel behavior of each commuter is determined. Refer to [6] for the details of the algorithm.

Next, the algorithm is validated with actual data. Data of the Beijing public transportation system from June 1 to June 30, 2015 are employed as an example.

TABLE 7.1 Extraction of Commuting Characteristics

Card ID	S_h	S_w	R_h	R_w	T_h	T_w	N_{day}	N_{route}	N_{stop}	N_{time}
43,207	20,152	365	00986	00958	19	29	5	2	3	3
41,610	217	102	6–2	10–6	14	31	2	2	4	2
32,558	1533	5503	00359	00359	15	35	24	21	33	11
32,664	265	723	57,300–60,366	60,366–00741	16	36	15	10	19	9
86,147	10,485	20,295	00012	00012	19	25	3	5	4	2

Notes: S_h *is the most frequent place-of-residence stop.* S_w *is the most frequent place-of-work stop.* R_h *is the most frequent route for the home-to-work trip.* R_w *is the most frequent route for the work-to-home trip.* T_h *is the most frequent departure time of the home-to-work trip.* T_w *is the most frequent departure time of the work-to-home trip.* N_{day} *is the number of travel days.* N_{route} *is the sum of the numbers of occurrences of the most frequent routes.* N_{stop} *is the sum of the numbers of the similarity of stops.* N_{time} *is the sum of the numbers of occurrences of the most frequent departure times.*

The data consist of subway data (107,481,320 items), urban bus data (215,144,507 items), and suburban bus data (42,220,547 items). Commuters are rated based on four indices: the number of travel days, same commuter routes, same commuter stops, and same departure times.

The algorithm identifies a total of 4,706,010 commuters. The extracted most frequent place-of-residence stops are projected onto a GIS map (shown as a heat map in Fig. 7.3).

The heat map in Fig. 7.3 visually depicts the distribution of places of residence of Beijing public transportation commuters. The dark color (red color) indicates a relatively high commuter population density, and the light color (blue color) indicates a relatively low commuter population density. As demonstrated in the heat map, most of the places of residence are distributed along the subway lines.

Some typical places of residence can also be observed on the heat map, e.g., the places of residence around Shahe Station and the Life Science Park station along the Changping line in the north, Tiantongyuan Station and Huilongguan

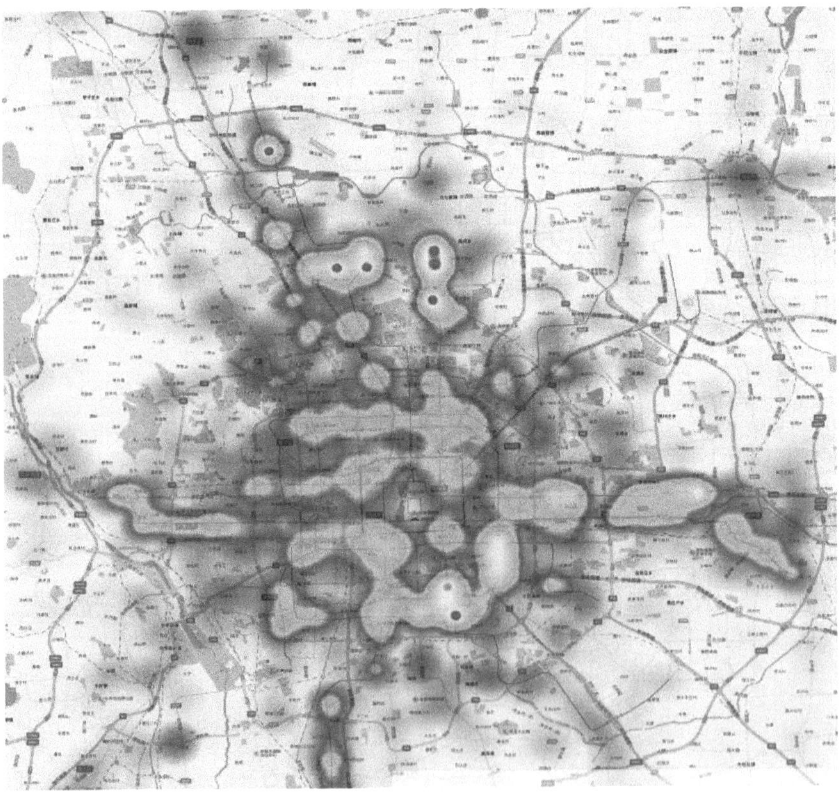

FIG. 7.3 Heat map of the places of residence of Beijing public transportation commuters in June 2015.

Station along Line 13, Puhuangyu Station and Liujiayao Station along Line 5 in the south, Shuangjing Station in the east and Caofang Station along Line 6. These places of residence are not only representative but also consistent with the actual situation. Fig. 7.4 displays the heat map of the extracted most frequent places of work.

The distribution of the places of work of Beijing public transportation commuters is visualized in the heat map in Fig. 7.4. Compared with the places of residence, the places of work in Beijing are concentrated and primarily distributed in the West 2nd Ring Road region, the East 2nd Ring Road region, the West 3rd Ring Road region, the Zhongguancun region, and the Xi'erqi region. The Central Business District has the highest place-of-work density, followed by Zhongguancun. Southern Beijing has a relatively low place-of-work distribution density, which is consistent with the actual situation.

To understand the distribution of commuters along the ring roads in Beijing, clustered stops with latitude and longitude coordinates are projected onto a GIS map. They are then classified into six groups based on the ring roads where they are located, as shown in Fig. 7.5: stops outside the 6th ring road; stops between the 6th ring road and the 5th ring road; stops between the 5th ring road and the

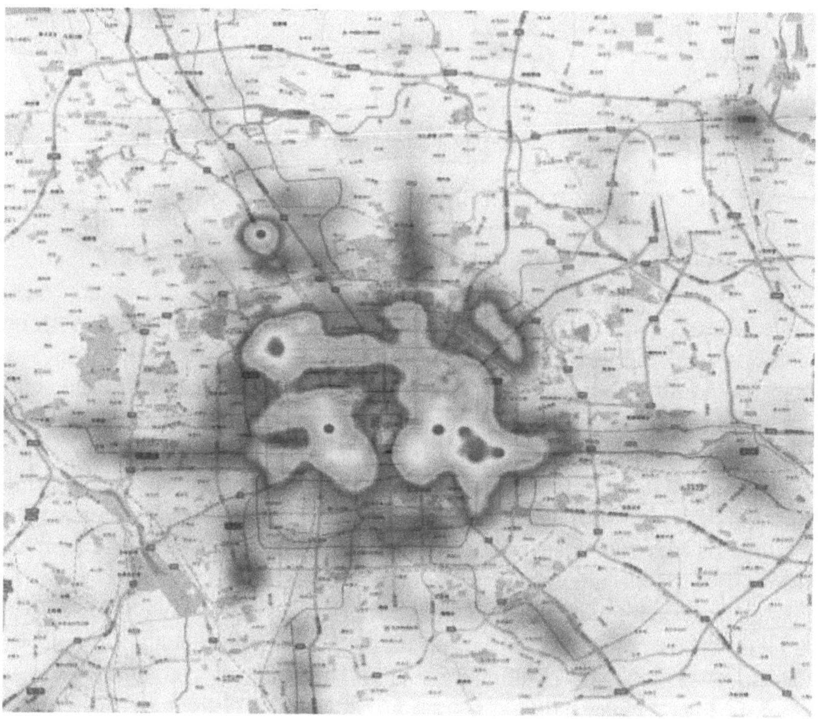

FIG. 7.4 Heat map of the places of work of Beijing public transport commuters in June 2015.

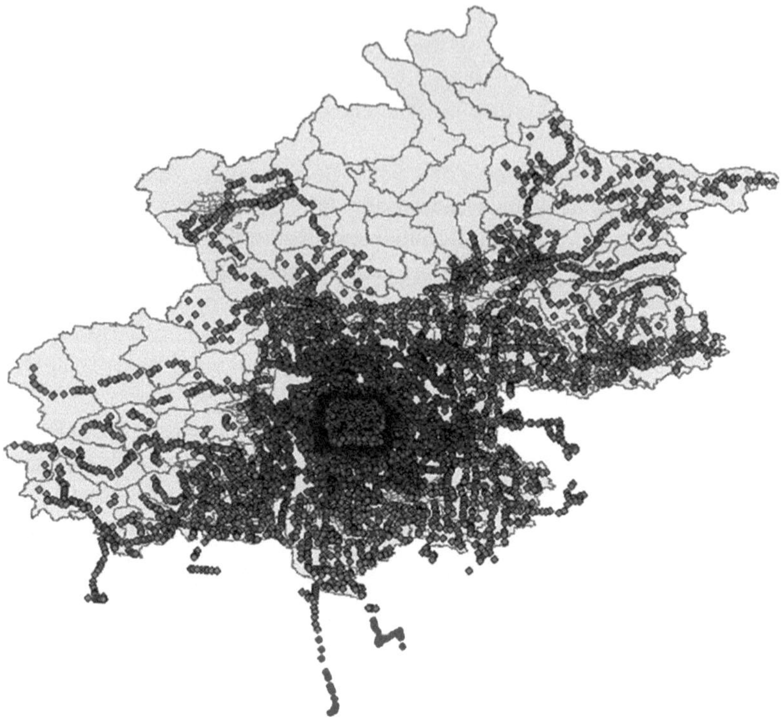

FIG. 7.5 Classification of stop IDs based on the ring roads where they are located.

4th ring road; stops between the 4th ring road and 3rd ring road; stops between the 3rd ring road and 2nd ring road; and stops within the 2nd ring road.

Based on the classification of stop IDs on the ring roads, the distributions of the places of residence and work of commuters can be analyzed. Table 7.2 lists the numbers of commuters at places of residence and work on each ring road and their percentage of the total.

As demonstrated in Table 7.2, most commuters have residences located outside the 3rd ring road, which accounts for approximately 83% of commuters. The 3rd ring road primarily contains busy commercial and administrative districts. Compared with other regions, the residence region between the 6th ring road and 5th ring roads has the largest number of commuters. The number of commuters with a residence outside the 5th ring road accounts for approximately 47% of the total number of commuters, which is attributed to lower housing prices or rents. Compared with the residences of commuters, most commuters work at places within the 6th ring road, which accounts for approximately 91% of all commuters. The number of commuters varies insignificantly between the work region between the 6th ring road and the 5th ring roads, the work region between the 5th ring road and the 4th ring road,

TABLE 7.2 Numbers of Commuters at Places of Residence and Work on Each Ring Road and Their Percentage of the Total

Ring Road	No. of Commuters at Resident Places	Percentage	No. of Commuters at Work Places	Percentage
Outside the 6th	576,471	12.42	391,572	8.43
6th–5th	1,638,422	35.31	821,710	17.68
5th–4th	783,918	16.90	725,858	15.62
4th–3rd	846,079	18.24	1,146,259	24.66
3rd–2nd	513,459	11.07	918,226	19.76
Within the 2nd	281,424	6.07	644,010	13.86

the work region between the 4th ring road and the 3rd ring road, the work region between the 3rd ring road and the 2nd ring road, and the work region within the 2nd ring road. In addition, most commuters (45% of all commuters) work in the region between the 4th ring road and the 2nd ring roads, where most of the commercial districts and places of work are distributed.

Based on the distribution of the places of work and residence of commuters along the ring roads, we can visually and comprehensively determine that the places of work are centrally distributed, whereas the places of residence are distributed along the subway lines. In addition, the places of work and residences of commuters are significantly unevenly distributed. These results can be used as data support for urban planning departments.

7.4 APPLICATION OF PUBLIC TRANSPORTATION DATA IN OPERATION AND MANAGEMENT

Informatization of an urban public transportation system can improve the effective utilization rate of public transportation vehicles to achieve dynamic scheduling and operation management of the public transportation system. Therefore, informatization of public transportation is important content for improving an urban public transportation system. For passengers, informatization can also provide tremendous convenience. By dynamically acquiring and analyzing the traveling information of public transportation vehicles and road and traffic information using advanced information and communication technologies and publishing the results via electronic public transportation stop boards, cell phones or the Internet in real time, information, such as the levels of public transportation congestion, travel time of public transportation vehicles and transfers, can be provided to passengers. This section introduces the application of public transportation data in the prediction of bus arrival times, i.e., the prediction of passengers' waiting times and their confidence intervals. The range of bus arrival times is introduced to quantify the uncertainty of prediction results, which can help passengers become aware of the operational information of public transportation and enable them to make reasonable adjustments to their travel time.

7.4.1 Prediction Model for Public Transportation Bus Arrival Times

This section introduces a smart card data-based prediction method for the range of public transportation bus arrival (passengers' waiting) times. This method predicts bus arrival times and their confidence intervals based on a relevance vector machine (RVM).

The RVM, which was proposed by Tipping in 2001, is a learning machine that was constructed based on a sparse Bayesian framework [7]. It is a supervised machine-learning algorithm that is often employed to solve regression and classification problems [8].

Next, variables that affect bus arrival times are selected based on smart card data. The following key fields are extracted from the smart card data: the datasets of smart card IDs, bus IDs, transaction times, route numbers, stop IDs, arrival times, dates, number of boarding passengers, and number of alighting passengers.

The arrival time of a bus at a certain stop is irregular and affected by a number of factors, such as the level of traffic congestion and the number of boarding and alighting passengers at each stop. The level of traffic congestion directly affects the travel time of a bus between two stops. The number of boarding and alighting passengers can affect the dwell time that a bus waits at a stop. The arrival time of a bus can be predicted by predicting the bus headway at the target stop. Bus headway is a measurement of the difference between the arrival times of two consecutive buses at a certain stop. If the time headway between two buses at the previous stop is known, the arrival time of the back bus at the following stop can be determined by predicting its time headway at the following stop. Therefore, the headway of a bus at the previous stop is an important factor that affects its arrival time at the following stop.

Stops A and B are assumed to be two stops on a bus route, and bus i and bus $i + 1$ (the bus of the same route following bus i) are assumed to pass stops A and B successively. Stops A and B can be two adjacent stops or two stops that are relatively far apart. Stop B is the target stop for the bus headway prediction. Thus, stop A is the upstream stop. In this section, the following six variables are selected as factors that affect bus arrival times, which are used as the input features of the model:

(1) bus headway at the upstream stop A, h_A;
(2) number of passengers that board on bus i at the upstream stop A, $b_{i,A}$;
(3) number of passengers that alight from bus i at the upstream stop A, $a_{i,A}$;
(4) number of passengers that board on bus $i+1$ at the upstream stop A, $b_{i+1,A}$;
(5) number of passengers that alight from bus $i+1$ at the upstream stop A, $a_{i+1,A}$; and
(6) travel time of bus i between the upstream stop A and the target stop B, $T_{i,A}T_{i,B}$.

The number of boarding and alighting passengers can be calculated based on the total number of smart card transactions that occur at each stop. The travel time of a bus between two stops is equal to the difference between the arrival times of the bus at the two stops, as shown in Eq. (7.1). In this equation, $T_{i,j-1}$ represents the arrival time of bus i at stop $j-1$, and $T_{i,j}$ represents the arrival time of bus i at stop j.

$$TT_{\imath,\jmath-1} - T_{i,j} - T_{i,j-1} \tag{7.1}$$

Then, the bus headway at target stop B (h_B) is the output variable of the model.

7.4.2 Case Study

Based on the previously mentioned analysis, six factors affect bus arrival times. These six factors are used as the input features of the RVM model. Refer to [9] for the details of the algorithm.

A case study of the Chang 51 route of the Beijing automatic fare collection (AFC) system is performed. Data of the Chang 51 route from July 1 to October 31, 2012 are selected, of which the data of the first 3 months are used to train the RVM model and the data of the last month is used to test the RVM model. To verify the validity and effectiveness of the RVM algorithm, it is compared with a support vector machine (SVM) algorithm. Two evaluation indices—mean absolute percentage error (MAPE) and root-mean-square error (RMSE)—are established as follows:

$$\text{MAPE} = \frac{1}{n}\sum_{i=1}^{n}\left|\frac{h(i)-\hat{h}(i)}{\overline{h}}\right| \times 100\% \tag{7.2}$$

$$\text{RMSE} = \sqrt{\frac{1}{n}\sum_{i=1}^{n}\left[h(i)-\hat{h}(i)\right]^2} \tag{7.3}$$

where $h(i)$ and $\hat{h}(i)$ are the real and predicted bus headway values at time i, respectively. Fig. 7.6 shows the comparison of the true values and the predicted values using the RVM and SVM algorithms. As demonstrated in Fig. 7.6, the predicted values that are obtained using the RVM algorithm are similar to the true values. As shown in Table 7.3, the MAPE and RMSE of the RVM

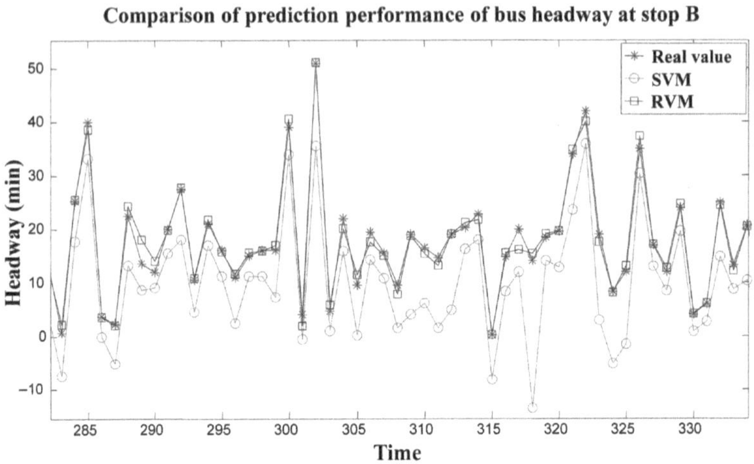

FIG. 7.6 Comparison of the true values and the predicted values that are obtained using the RVM and SVM algorithms.

TABLE 7.3 Errors of the RVM and SVM Algorithm

Algorithm	MAPE (%)	RMSE
RVM	14.63	3.5147
SVM	35.76	8.8054

algorithm are less than the MAPE and RMSE of the SVM algorithm, which demonstrates the validity of the RVM algorithm.

Fig. 7.7 shows the prediction results of the confidence interval of the bus headway based on the RVM algorithm (confidence level = 95%). As demonstrated in this figure, most of the real values fall within the confidence interval (RVM- and RVM+ represents the lower and upper bound of the confidence interval of the predicted values, respectively). The mean value of the actual bus headways is 18 min, whereas the width of the confidence interval is approximately 7 min, which indicates that passengers only need to arrive at the bus stop 7 min earlier. This interval will help improve passengers' travel experience. Thus, the effectiveness of the RVM algorithm is verified.

7.5 INTRODUCTION OF A PUBLIC TRANSPORTATION BIG DATA PLATFORM BASED ON *E*-SCIENCE

The previously mentioned analysis that relates to public transportation big data is important for improving public transportation service quality, optimizing the public transportation planning strategy, and motivating people to use public

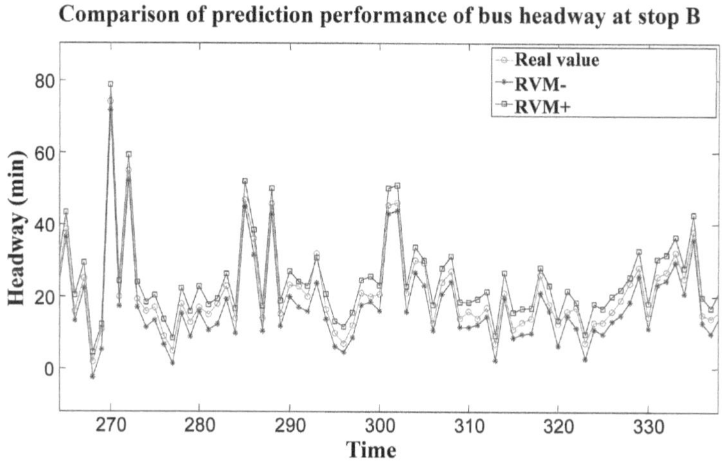

FIG. 7.7 Comparison of the confidence interval of the predicted values that are obtained using the RVM algorithm and the true values.

transportation. In addition, the previously mentioned analysis is crucial to the reasonable evaluation of the operating conditions of the public transportation network. An advanced public transportation network performance evaluation system can enable the public transportation department to make decisions and determine when, where, and how to provide better transport services. In addition, this system not only can facilitate public transportation departments to produce technical summaries and perform statistical analysis of the performance and operating conditions of public transportation networks, but also can make passengers aware of the operating conditions of urban transportation networks by publishing relevant information. Based on these reasons, an *E*-Science platform for public transportation is constructed in this section based on the theoretical framework of *E*-Science to evaluate the performance of public transportation networks.

E-Science was proposed in the United Kingdom as a brand new mode of scientific research that was established based on Internet technology and wide-area distributed high-performance computing environments to address the unprecedentedly complex problems in various disciplines and research fields, i.e., scientific research activities in support of information infrastructure.

Thus, this section primarily introduces a public transportation big data platform that is based on the theoretical framework of *E*-Science [10], which was established using a combination of spatial data (geospatial data of public transportation network) and public transportation data (public transportation smart card data)—an *E*-Science public transportation performance evaluation platform. This platform is an *E*-Science platform with integrated public transportation data sharing, visualization, modeling, and analysis functions. This platform can calculate public transportation performance indices, evaluate the performance of a public transportation network, and display the results to users.

7.5.1 Main Functions of the Public Transportation Big Data Platform

The Beijing AFC system and automated vehicle location (AVL) system data are the two main data sources for this platform. The Beijing AFC system entered service in the public transportation system on May 10, 2006. This platform provides various performance evaluation indices for public transportation [11]. Performance evaluation indices of various levels for public transportation are introduced as follows.

7.5.1.1 Network-Level Evaluation Index for Operating Speed

A minimum travel time-based transport route search algorithm that employs real-time traffic speed data is used to enable travelers to plan routes and formulate travel plans. By observing low-speed routes, public transit agencies can also identify congestion bottlenecks and study the spreading process of traffic congestion over networks. Then, public transit agencies can mitigate traffic congestion by relevant policies (e.g., transport route optimization). Smart card transaction data include time information (transaction time) and spatial information (public transportation routes and O and D information). Based on smart card transaction data, the travel time between two stops can be estimated. The travel time between two adjacent stops is composed of operating time and delay time. Delay time refers to the length of time that a bus waits at a certain stop to load and unload passengers, which is highly correlated to the number of passengers who wait at the stop. Therefore, when calculating the actual operating speed, the general traffic conditions and the delay time caused by passengers should be considered. Thus, the average speed of a bus between two stops can be calculated by dividing the network distance between the two stops by the travel time of the bus between the two stops. The network-level evaluation index for traffic speed can provide bus drivers with dynamic route information and public transportation departments with optimized public transportation networks and public transportation researchers with traffic congestion diagnoses.

7.5.1.2 Route-Level Reliability Evaluation Index for Public Transportation Travel Time

Travel time reliability represents the temporal stability of repeated trips. Providing travel time reliability information can help passengers plan their trips and reduce traffic congestion. In public transportation, public transportation travel time reliability affects the attractiveness and efficiency of transport services and is related to the on-time performance, which is indicative of customers' satisfaction and the stop-level headway variance. From the passengers' perspective, their travel time encompasses on-board travel time and waiting time. Route-level transport travel time reliability can be used to measure the variance of on-board travel time. Numerous effective methods are available for quantifying transport travel time reliability.

(1) 90th or 95th percentile travel time: this describes travel time during heavy traffic conditions.
(2) Buffer index: this calculates the time spent by a passenger in addition to the average travel time to ensure an on-time arrival. Additional time can be defined as the difference between the 95th percentile travel time and the average travel time. Thus, the buffer index is the ratio of the additional time to the average travel time.
(3) Scheduled time index: this is the ratio of the 95th percentile travel time to the travel time in smooth traffic conditions.

7.5.1.3 Stop-Level Ridership Evaluation Index

The stop-level ridership at each stop along a route refers to the number of boarding and alighting passengers at the respective stop. Ridership serves an important role in facilitating the public transit agencies to monitor transport services and attain economic benefits. Transport operators can identify the stops where with a large number of boarding passengers based on the stop-level ridership and adjust the transit schedules to improve public transportation service quality. Public transportation decision makers can evaluate the effectiveness of a new fare policy and respond to fare changes based on the public transportation ridership. The total number of boarding and alighting passengers determines the passenger transport demand of each route and affects the marketing and operational strategies of public transportation departments.

7.5.1.4 Stop-Level Headway Variance

Bus headway is a key factor that measures transport service reliability; it can be defined as the difference between the arrival times of two consecutive buses at a certain stop along the same route. Bus drivers and passengers prefer constant bus headways. A small bus headway will cause bus bunching, whereas a large bus headway will produce an increase in passengers' waiting time. Irregular headways will tarnish the attractiveness of public transportation. Because changes in traffic signals and the number of boarding and alighting passengers will cause changes in the bus headways at different stops, for transit schedules, stop-level headways must be analyzed to adjust the headways in the middle of the route.

7.5.2 Functions of the Public Transportation Big Data Platform

The four core components of this platform are as follows.

7.5.2.1 Public Transportation Network Speed Map

The travel speed of public transportation is calculated based on the identified passenger O-D pairs. Fig. 7.8 shows the traffic conditions of the Beijing public transportation network between 4:30 p.m. and 5:00 p.m. on July 31, 2010 (a workday).

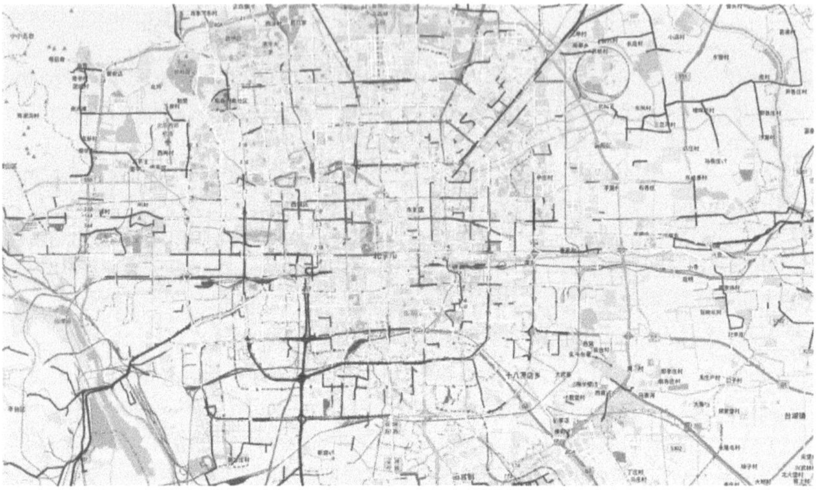

FIG. 7.8 Beijing public transportation network speed map. *(Map data sources:* Open Street Map.*)*

This platform provides an additional analysis function for checking the statistical data of the travel speed of public transportation within the network. A window that shows detailed traffic speed information about the entire network (e.g., average speed, variance, the 90th percentile speed, percentage of traffic sections that are congested/uncongested, composition of data sources for bus speed calculation (GPS and smart card data)) will pop up when the operating speed statistics button is clicked. Public transportation departments can use the color speed map and statistical analysis as an effective tool for identifying congested areas and make corresponding improvements in its public transportation services, e.g., opening high-speed roads or reducing headway for public transportation vehicles.

7.5.2.2 Public Transportation Ridership Analysis

Public transportation ridership analysis is important to public transportation departments. They strive to retain existing public transportation passengers and attract potential public transportation passengers. In addition, the spatial and temporal public transportation ridership shown on a map system can facilitate public transportation departments to perform a comparative analysis and evaluate changes in passengers' demand. As demonstrated in Fig. 7.9, the radius of each stop indicates the magnitude of the passenger load. The passenger load at each stop is defined as the difference between the number of boarding passengers and the number of alighting passengers.

7.5.2.3 Distribution of Bus Headways

Fig. 7.10 shows the bus headway of the No. 118 bus route on April 7, 2008. Estimation based on the smart card transaction data indicates a total of 116

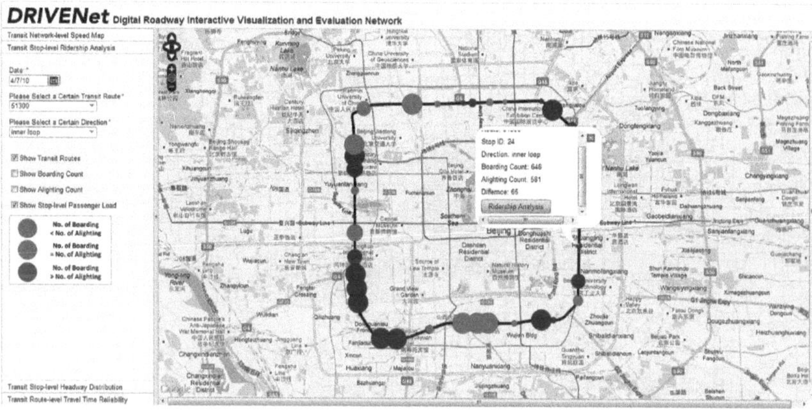

FIG. 7.9 Analysis of the ridership of route 51,300. *(Map data sources: Google and AutoNavi.)*

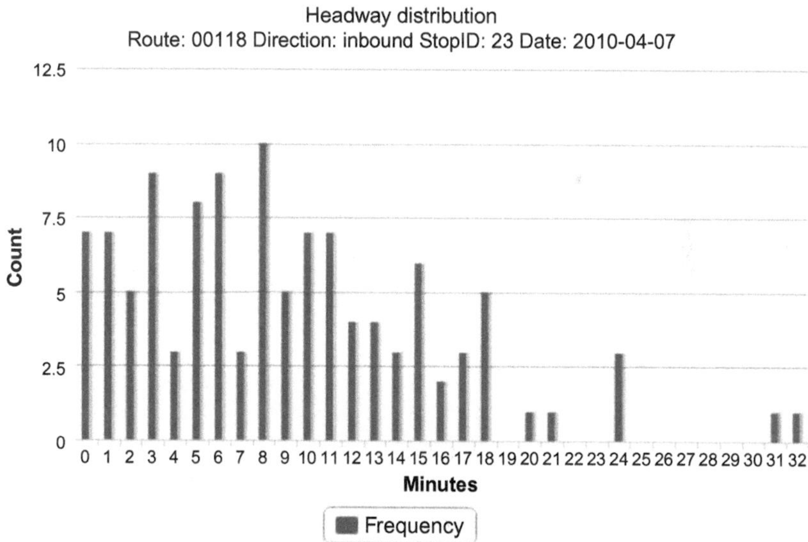

FIG. 7.10 A histogram of bus headways at a particular bus stop. *(Map data sources: Google and AutoNavi.)*

bus trips. The average bus headway and variance of bus headways at 23 stops can be calculated (9.47 min and 6.64 min, respectively). A frequency histogram is generated to display the distribution of headways.

7.5.2.4 Public Transportation Travel Time Reliability

Another function of this platform is to analyze public transportation travel time reliability based on a GIS map, as shown in Fig. 7.11A. The buffer time index is used to measure travel time reliability. The smaller the buffer time index is, the more reliable the public transportation route is. To analyze changes in the public

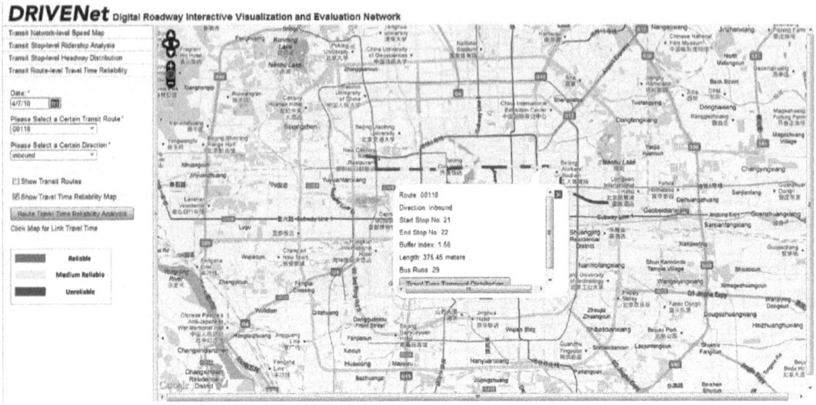

(A)

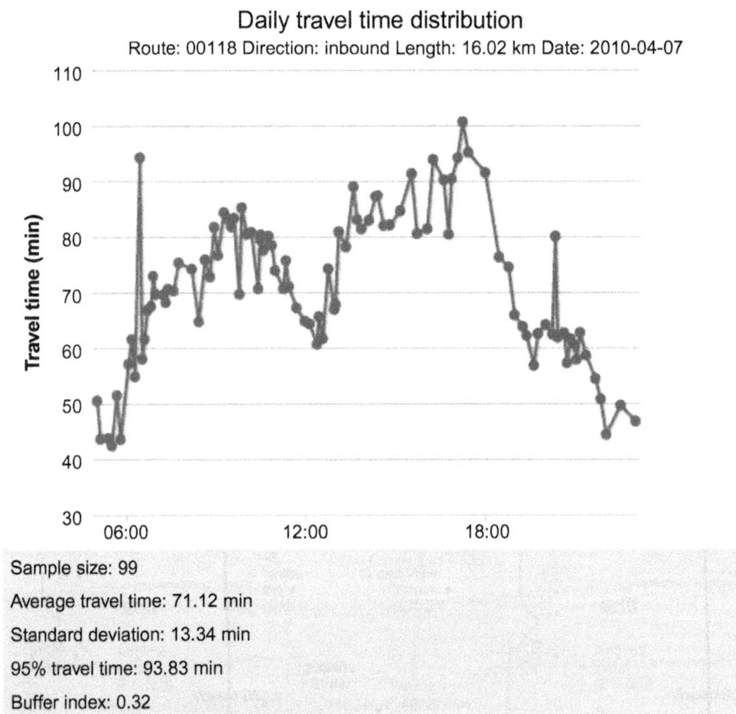

(B)

FIG. 7.11 (A) Spatial distribution of bus travel time reliability; (B) trend analysis of bus travel time. *(Map data sources: Google and AutoNavi.)*

transportation travel time, the travel time on each passing route/link can be calculated, as demonstrated in Fig. 7.11B. The buffer time index of the No. 118 bus route on April 7, 2010 is 0.32, which indicates that on-time arrivals with 95% can be ensured once bus passengers have waited 22.71 min.

7.6 CONCLUSIONS

In the big data era, a central issue is how to obtain correct, noise-free effective data from massive intelligent public transportation system data using efficient data processing methods, extract public transportation ridership, and vehicle information from these effective data. The extracted information then needs to be deeply mined to ensure that relevant analysis and prediction results are applicable to public transportation operation, planning, and management.

This chapter introduced the sources and types of data generated by public transportation systems against a big data background, as well as preprocessing methods for public transportation smart card and GPS data. Then, the application of big data in public transportation planning, operation, and management, and public transportation network evaluation was introduced. The validity and feasibility of each method proposed in this chapter was verified by a case study using Beijing public transportation system data. This chapter focused on transport data mining and analysis; the results can be used to provide public transportation information services to the public, reduce passengers' travel times on public transportation, improve the operational and management level of public transportation, and provide data support to the public transportation management department in their decision-making process.

ACKNOWLEDGMENT

This chapter is supported by the National Natural Science Foundation of China (61773036), Beijing Natural Science Foundation (9172011), and Young Elite Scientist Sponsorship Program by the China Association for Science and Technology (2016QNRC001).

REFERENCES

[1] M. Pelletier, M. Trépanier, C. Morency, Smart card data use in public transit: a literature review, Transp. Res. C Emerg. Technol. 19 (4) (2011) 557–568.

[2] S. Robinson, B. Narayanan, N. Toh, F. Pereira, Methods for pre-processing smartcard data to improve data quality, Transp. Res. C Emerg. Technol. 49 (2014) 43–58.

[3] X. Ma, C. Liu, J. Liu, F. Chen, H. Yu, Boarding stop inference based on transit IC card data (in Chinese), J. Transp. Syst. Eng. Inf. Technol. 15 (4) (2015) 78–84.

[4] G. Ball, J. Hall, A Novel Method of Data Analysis and Pattern Classification, Stanford Research Institute, 1965.

[5] C. Hwang, K. Yoon, Multiple attribute decision making: methods and applications, Lecture Notes Econ. Math. Syst. 375 (4) (1981) 59–191.

[6] X. Ma, C. Liu, H. Wen, Y. Wang, Y. Wu, Understanding commuting patterns using transit smart card data, J. Transp. Geogr. 58 (2017) 135–145.
[7] J. Lou, Y. Jiang, Q. Shen, Z. Shen, Z. Wang, R. Wang, Software reliability prediction via relevance vector regression, Neurocomputing 186 (C) (2016) 66–73.
[8] M. Tipping, Sparse Bayesian learning and the relevance vector machine, J. Mach. Learn. Res. 1 (3) (2001) 211–244.
[9] H. Yu, Z. Wu, D. Chen, X. Ma, Probabilistic prediction of bus headway using relevance vector machine regression, IEEE Trans. Intell Transp. Syst. 18 (7) (2017) 1772–1781.
[10] X. Ma, Y. Wu, Y. Wang, DRIVE net: an E-science of transportation platform for data sharing, visualization, modeling, and analysis, Transp. Res. Rec. 2215 (2011) 37–49.
[11] X. Ma, Y. Wang, Development of a data-driven platform for transit performance measures using smart card data and GPS data, J. Transp. Eng. 140 (12) (2014).

Chapter 8

Simulation-Based Optimization for Network Modeling With Heterogeneous Data

Xiqun (Michael) Chen

College of Civil Engineering and Architecture, Zhejiang University, Hangzhou, People's Republic of China

Chapter Outline

8.1 INTRODUCTION

This chapter introduces simulation-based optimization (SBO) methodologies for real-world large-scale network modeling with heterogeneous data. The framework presented in this chapter is primarily based on the author's previous work [1–7]. This chapter can be referred to as a tutorial of SBO procedures for solving computationally costly transportation optimization problems, in particular, in large-scale networks with complicated traffic dynamics.

To evaluate the transportation system performance under certain planning and operations strategies, the influence of implemented measures on the network characteristics and the resulting impact on travelers' behavior need to be modeled. At the macroscopic level, travelers' behavior is usually modeled by mathematical equations with various assumptions (e.g., Wardrops' first principle, and utility maximization). These mathematical equations have been

Data-Driven Solutions to Transportation Problems. https://doi.org/10.1016/B978-0-12-817026-7.00009-6

applied in the widely used four-step model as well as the more advanced activity-based travel demand models. On the other hand, simulation models have been developed for the mesoscopic/microscopic-level analysis of travel behavior. Simulation has a better capability in describing the dynamics of travel behavior than typical analytical models.

However, in a real-world large-scale transportation network, searching for the optimal solution usually requires intensive objective function evaluations. The objectives are usually expensive to evaluate especially when using black-box simulations to assess planning policies and/or operational strategies. A single function evaluation may require several hours to several days of complete simulations. In this case, it may be necessary to forgo an exact evaluation and instead use an approximate response surface that is computationally efficient. Surrogate models provide a promising approach to solving network optimization problems.

From different perspectives, the objectives can be minimizing the average trip travel time of the whole network, maximizing the vehicle throughput, and/or toll revenue, etc. Consider the following continuous simulation optimization problem:

$$\min_{\mathbf{x}\in\Omega} f(\mathbf{x}) = \mathrm{E}[F(\mathbf{x}, \boldsymbol{\omega})] \tag{8.1}$$

where $\mathbf{x}$ is a k-dimensional vector of continuous decision variables, and Ω is the parameter space. The objective function $f(\mathbf{x})$ has no analytically closed-form or differentiability, and can only be evaluated by black-box stochastic simulation. The stochastic response $F(\mathbf{x}, \boldsymbol{\omega})$ has inputs of the design variables $\mathbf{x}$ and a random sample from the distribution of $\boldsymbol{\omega}$, and $\boldsymbol{\omega}$ is a random variable defined in the probability space $(\mathcal{W}, \mathcal{F}, \mathcal{P})$. The response function $F(\mathbf{x}, \boldsymbol{\omega})$ is measurable so that the expectation $\mathrm{E}[F(\mathbf{x}, \boldsymbol{\omega})]$ exists for every $\mathbf{x}$. In what follows, $\boldsymbol{\omega}$ is suppressed in $F(\mathbf{x}, \boldsymbol{\omega})$ for ease of presentation. The SBO approach aims to search for the optimal decision variables $\mathbf{x}^*$. The optimization problem (8.1) has three important attributes:

(1) system dynamics are explicitly simulated by the individual travel behaviors and can't be represented by a set of equalities and inequalities;
(2) the objective function is expensive to evaluate because every evaluation requires one or more model runs for each set of decision variable values, while each run of simulation usually takes a few hours;
(3) the optimization problem has no closed-form objectives and usually has an implicit expression of decision variables when incorporating large-scale transportation simulations;
(4) it is impractical to accurately approximate the derivatives of the objective function. Thus, traditional optimization approaches are not applicable to solve such a problem.

To bridge this gap, this chapter adopts the SBO approach that reconstructs the response surface by only a few evaluations of the objective function and is capable of handling simulation noises. This chapter aims to provide a macroscopic understanding of urban traffic dynamics by using both a simulation-based dynamic traffic assignment model and heterogeneous traffic detection data. This chapter validates the mesoscopic simulation generated macroscopic fundamental diagrams (MFDs) by using fixed freeway traffic flow detectors, and reveals empirical relationships between link-based (time aggregate) and path-based (individual traveler and time disaggregate) network-wide traffic flow statistics. Finally, the SBO approach is demonstrated in the congestion pricing of a real-world large-scale transportation network.

The rest of the chapter is organized as follows: Section 8.2 reviews the related SBO literature. Section 8.3 proposes a hierarchical framework of transportation SBO. Section 8.4 demonstrates the field application of the SBO approach, where the simulation model is validated by heterogeneous empirical data including fixed loop/microwave traffic flow measurements on freeways and probe vehicle based travel time observations in major arterials. Finally, Section 8.5 concludes the chapter.

8.2 LITERATURE REVIEW

SBO approaches that require fewer evaluations of the objective function have been proven to be powerful and effective in a broad of optimization problems, e.g., congestion pricing [1], mixed road pricing policies [2, 3], mixed integer network design problem [4], improving network-wide travel time reliability [5], optimization of time-of-day vehicle mileage fees [6], multiobjective time-varying pricing [7], traffic signal control problem [8–12], sustainable transportation systems evaluation and optimization [13], transportation corridor active traffic management [14], social welfare maximization of multimodal transportation systems [15], joint optimization of travel behavioral responses and traffic management with the Bayesian stochastic Kriging metamodel [16] and bootstrapped support vector regression [17], design and evaluation of flexible work schedules [18], optimal en-route travel information provision [19], transportation supply parameter calibration [20], transportation demand parameter calibration [21], and network design problems [22].

Surrogate modeling or metamodels aim to regress the response surface that characterizes the relationship between input decision variables and simulation outputs [23–25]. The surrogate simplifies simulation optimization because of the deterministic rather than stochastic relationship between the input and output [26]. Using only an initial input dataset and corresponding output values of the objective function, the surrogate model can be developed as an approximation of the expensive-to-evaluate objective function. An unknown function that formulates the relationship between simulation input and output is approximated with a predefined parametric function whose coefficients can be determined via the experiment design [27]. On the basis of exploration and exploitation of the computationally efficient surrogate, the optimal solution can be obtained. One smooth continuous approximation function was unable to fit the discontinuous simulation output due to random noises around the true average response value, while the optimization accuracy relied on how accurate the surrogate models were in capturing the performance variations [28].

Validation of a real-world network remains a challenge for both traffic researchers and practitioners. With an efficient monitoring of large-scale networks using heterogeneous sources of data, e.g., fixed loop detectors and probe vehicle sensors, time-varying network performance measures can be applicable, e.g., average network speed, network travel time of all drivers, or statistics based on individual trip travel times.

The MFD, which is also known as the network fundamental diagram (NFD), has been widely used to evaluate the macroscopic traffic state dynamic for two-dimensional urban road networks or freeway networks. It is a property of the road network under certain traffic control strategies but not of travel demand. Research on network-wide traffic flow relationships can be tracked back to the 1960s [29]. The properties of MFD or NFD were further investigated using microsimulation and real-world data [30]. A well-defined MFD was observed existing between space-mean flow and density by spatially aggregating the highly scattered network-wide statistics from individual loop detectors in homogeneous networks [31]. Recently, numerous empirical and microsimulation studies observed the existence of a reproducible and well-defined relation between network average flow and density for urban and freeway networks [32–35]. These studies link the details in individual links with the description of the congestion level of networks and its dynamics.

As transportation simulations take account of the interactions between complex travel demand patterns and network supply, it is always time-consuming to evaluate system performance. Thus, an SBO method that requires fewer objective function evaluations is desired to solve the optimization problem. In this chapter, the toll policy for a freeway sub-network is optimized to change the spatial distribution of congestion such that a number of objectives can be achieved, e.g., average travel time decreases, network outflow increases, toll revenue increases, and simultaneous consideration of aforementioned goals.

8.3 SIMULATION-BASED OPTIMIZATION

8.3.1 Framework

This section presents the SBO framework to solve the expensive-to-evaluate large-scale network modeling problems. This approach is computationally efficient and suitable for complex problems with very tight computational budgets. A systematic SBO framework is presented in Fig. 8.1. There are four main components: network and demand data, simulation, model validation, and simulated-based optimization. OD matrices are constructed based on travel demands of a regional planning model [36]. In this chapter, the objectives include the average

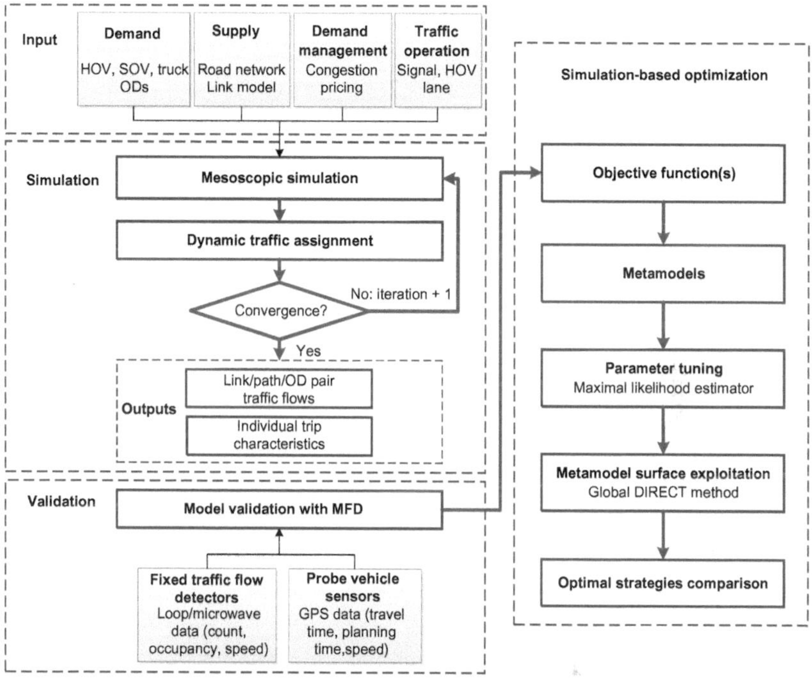

FIG. 8.1 A systematic SBO framework for network modeling with heterogeneous data.

travel time minimization, vehicle throughput maximization, toll revenue maximization, and overall desirability maximization. The mesoscopic simulation-based DTA model, DynusT [37], is incorporated to reveal traffic flow dynamics, based on which the network performance can be evaluated.

8.3.2 Design of Experiments (DoE)

Given the objective function evaluated by simulation outputs, an initial number of strategies are generated through the DoE, e.g., Latin Hypercube Sampling (LHS). A space filling DoE is useful when only a few runs of simulation can be afforded within the computational budget. In the LHS, each dimension of the decision variable is split into a relatively large number of equal-size bins, in which subsamples are uniformly generated. The advantage of LHS is that the mappings of high dimensional design inputs into each dimension are uniformly distributed without overlap. This property makes LHS one of the space-filling types of DoE.

8.3.3 Surrogate Models

8.3.3.1 Quadratic Polynomial Function (QPF)

Initially, there are n strategies that form a sampling plan $\mathbf{X} = [\mathbf{x}^{(1)}, \mathbf{x}^{(2)}, \ldots, \mathbf{x}^{(n)}]^{\mathrm{T}}$ and responses $\mathbf{y} = [y^{(1)}, y^{(2)}, \ldots, y^{(n)}]^{\mathrm{T}}$, where $\mathbf{x}^{(i)}$ denotes the ith strategy.

Let $\hat{f}(\mathbf{x})$ be the estimate of the real objective function $f(\mathbf{x})$. The most commonly used surrogate model is the quadratic polynomial function given by:

$$\hat{f}(\mathbf{x}) = \beta_0 + \sum_{i=1}^{k} \beta_i x_i + \sum_{i<j}\sum_{j} \beta_{ij} x_i x_j + \sum_{i=1}^{k} \beta_{ii} x_i^2, \quad \mathbf{x} \in \mathbb{R}^k \tag{8.2}$$

where $\mathbf{x}$ is the new k-dimensional decision vector in Eq. (8.2), β_0 is the intercept, β_i are the linear coefficients, β_{ij} are the coefficients of interaction terms, β_{ii} are the quadratic coefficients.

8.3.3.2 *Radial Basis Function (RBF)*

RBF adopts the basis function $\varphi(r)$ that only depends on the radial distance r between $\mathbf{x}$ and each sample point $\mathbf{x}^{(i)}$. The RBF prediction at a new point is given by:

$$\hat{f}(\mathbf{x}) = \left(\mathbf{\Phi}^{-1}\mathbf{y}\right)^{\mathrm{T}} \boldsymbol{\varphi} \tag{8.3}$$

where $\mathbf{\Phi}$ is the Gram matrix, each element of which is $\mathbf{\Phi}_{i,j} = \varphi(\|\mathbf{x}^{(j)} - \mathbf{x}^{(i)}\|)$, $\boldsymbol{\varphi}$ is the Gaussian basis function, i.e., $\varphi(\|\mathbf{x} - \mathbf{x}^{(i)}\|) = \exp\left(-c\|\mathbf{x} - \mathbf{x}^{(i)}\|^2\right)$, $c > 0$ is the shape parameter, $\|\mathbf{x} - \mathbf{x}^{(i)}\|$ is the Euclidean norm.

8.3.3.3 *Kriging Models: Noise Free and Noise*

The Kriging model predicts a response by summarizing a linear model and a high-frequency variation component. $\boldsymbol{\theta}$ can be estimated by maximizing the likelihood of the observed data $\mathbf{y}$, with the likelihood given by:

$$L = \frac{1}{(2\pi\sigma^2)^{n/2}\sqrt{|\mathbf{\Psi}|}} \exp\left(\frac{(\mathbf{y} - \mathbf{1}\mu)^{\mathrm{T}}\mathbf{\Psi}^{-1}(\mathbf{y} - \mathbf{1}\mu)}{2\sigma^2}\right) \tag{8.4}$$

where μ is the mean of the objective function, σ^2 is the variance, $\mathbf{1}$ is the $n \times 1$ unit column, $\mathbf{\Psi}$ is the $n \times n$ variation correlation matrix with the element $\psi\left(\mathbf{x}^{(i)}, \mathbf{x}^{(j)}\right) = \exp\left(-\sum_{l=1}^{k} \theta_l \left(x_l^{(i)} - x_l^{(j)}\right)^2\right)$, $i, j = 1, \ldots, n$, $\boldsymbol{\psi} = [\psi(\mathbf{x}, \mathbf{x}^{(1)}), \ldots$ $\psi(\mathbf{x}, \mathbf{x}^{(n)})]^{\mathrm{T}}$ is a vector of correlations between the observed data and the new prediction, let $\boldsymbol{\theta} = [\theta_1, \ldots, \theta_k]^{\mathrm{T}}$ be with weights of each dimension, and the use of variable $\boldsymbol{\theta}$ gives an asymmetric basis, allowing for varying impacts of each variable of the design space.

The maximization of the likelihood function is simplified by taking the natural logarithm and ignoring constant terms, given by:

$$\log(L) = -\frac{n}{2}\log\left(\sigma^2\right) - \frac{n}{2}\log(|\mathbf{\Psi}|) - \frac{(\mathbf{y} - \mathbf{1}\mu)^{\mathrm{T}}\mathbf{\Psi}^{-1}(\mathbf{y} - \mathbf{1}\mu)}{2\sigma^2} + \text{constant} \tag{8.5}$$

Let the first derivatives of log(L) with respect to μ and σ^2 be zero, the maximum likelihood estimates are given by:

$$\hat{\mu} = \frac{\mathbf{1}^{\mathrm{T}}\boldsymbol{\Psi}^{-1}\mathbf{y}}{\mathbf{1}^{\mathrm{T}}\boldsymbol{\Psi}^{-1}\mathbf{1}} \tag{8.6}$$

$$\hat{\sigma}^2 = \frac{(\mathbf{y}-\mathbf{1}\hat{\mu})^{\mathrm{T}}\boldsymbol{\Psi}^{-1}(\mathbf{y}-\mathbf{1}\hat{\mu})}{n} \tag{8.7}$$

The concentrated log likelihood is given by:

$$-\frac{n}{2}\log\left(\hat{\sigma}^2\right) - \frac{n}{2}\log\left(|\boldsymbol{\Psi}|\right) \tag{8.8}$$

The maximum likelihood estimate of the objective function $f(\mathbf{x})$ by summarizing a linear model and a high-frequency variation component is given by:

$$\hat{f}(\mathbf{x}) = \hat{\mu} + \boldsymbol{\psi}^{\mathrm{T}}\boldsymbol{\Psi}^{-1}(\mathbf{y}-\mathbf{1}\hat{\mu}) \tag{8.9}$$

The error is now given by:

$$\hat{s}^2(\mathbf{x}) = \hat{\sigma}^2\left[1 - \boldsymbol{\psi}^{\mathrm{T}}\boldsymbol{\Psi}^{-1}\boldsymbol{\psi}\right] \tag{8.10}$$

The optimal $\hat{\boldsymbol{\theta}}$ is gained via maximizing the concentrated log likelihood function, and the global optima are obtained as $\mathbf{x}^* \in \underset{\mathbf{x}_{\min}\leq\mathbf{x}\leq\mathbf{x}_{\max}}{\arg\min}\ \hat{f}(\mathbf{x})$. The calibration and exploitation method will be introduced in Section 8.3.5.

The regressing Kriging model can be used by adding the error estimation of the observed data to the diagonal of the correlation matrix, so that the new correlation matrix is $\widetilde{\boldsymbol{\Psi}} = \boldsymbol{\Psi} + \lambda\mathbf{I}$, where λ is the regularization constant, and $\mathbf{I}$ is an identity matrix. So the optima given by the regressing Kriging model with noise can be more robust than the noise-free model.

8.3.4 Link-Based and Path-Based MFD

The validation of MFD aims to provide a macroscopic understanding of urban traffic dynamics. For the link-based MFD, to implement the macroscopic traffic analysis on the network level, we investigate the relationship between the accumulation of vehicles in a network with the exit outflows, and the equivalent relationship between the network-wide weighted average density and flow rate, given by:

$$N_t = \sum_{a\in A} k_{a,t} l_a \lambda_a, \quad N_t^{\text{freeway}} = \sum_{a\in\bar{A}} k_{a,t} l_a \lambda_a \tag{8.11}$$

where N_t is the time-varying number of vehicles in a network denoted by A, each individual link is $a \in A$, $k_{a,t}$ is the traffic density of link a at time t, l_a, and λ_a are the length and the number of lanes of link a. N_t^{freeway} is the time-varying number of vehicles in the freeway network denoted by $\overline{A} \in A$. For aggregate values, individual segment readings are weighted by their lane-length, i.e.:

$$K_t = \frac{N_t}{L} = \frac{\sum_{a \in A} k_{a,t} l_a \lambda_a}{\sum_{a \in A} l_a \lambda_a}, \quad K_t^{\text{freeway}} = \frac{N_t^{\text{freeway}}}{L} = \frac{\sum_{a \in A} k_{a,t} l_a \lambda_a}{\sum_{a \in \overline{A}} l_a \lambda_a} \tag{8.12}$$

where K_t is the space mean density (vehicle per mile) at time t, K_t^{freeway} is the space mean freeway density (vehicle per mile) at time t, L is the total length (lane-miles) of the freeway network. Analogously:

$$Q_t = \frac{\sum_{a \in A} q_{a,t} l_a \lambda_a}{\sum_{a \in A} l_a \lambda_a}, \quad Q_t^{\text{freeway}} = \frac{\sum_{a \in A} q_{a,t} l_a \lambda_a}{\sum_{a \in \overline{A}} l_a \lambda_a} \tag{8.13}$$

where Q_t is the space mean flow rate (vehicle per hour) of the network, Q_t^{freeway} is the space mean flow rate (vehicle per hour) of the freeway network, $q_{a,\,t}$ is the traffic flow rate of link a at time t.

According to the traffic variables relationship in the MFD, the freeway network average speed is estimated by:

$$V_t^{\text{freeway}} = \frac{Q_t^{\text{freeway}}}{K_t^{\text{freeway}}} \tag{8.14}$$

Historic average speeds for different time of day and day of week were used to estimate the arterial network mean average speed, given by:

$$V_t^{\text{arterial}} = \frac{\sum_{a \in \overline{A}^C} v_{a,t} l_a \lambda_a}{\sum_{a \in \overline{A}^C} l_a \lambda_a} \tag{8.15}$$

where V_t^{arterial} is the time-varying arterial average speed in the arterial network denoted by $\overline{A}^C$, which is the complement of set $\overline{A}$, $v_{a,\,t}$ is the traffic density of link a at time t, l_a, and λ_a are the length and the number of lanes of road segment $a \in \overline{A}^C$.

For the path-based MFD, in general, this theoretical linear relationship between the mean travel time rate and its standard deviation (SD) is given by:

$$\sigma_\tau = \theta_1 \overline{\tau} + \theta_2 \tag{8.16}$$

where $\overline{\tau}$ is average travel time rate (travel time per distance), σ_τ is the SD of the travel time rate, and θ_1, θ_2 are coefficients.

8.3.5 Calibration and Exploitation

The DIRECT (shortening of DIviding RECTangles) optimization algorithm was first proposed in [38] and motivated by a modification to Lipschitzian optimization to obtain a deterministic solution in SBO calibration and exploitation. DIRECT is a sampling algorithm that requires no knowledge of the objective function's gradient. It does not try to exploit any correlation between the function value at one point and its value at nearby points, and thus can be very useful when the objective function is a black box function or simulation. The DIRECT algorithm will globally converge to the global minimum, though the convergence may come at the expense of an exhaustive search over the domain. Since the algorithm needs dense evaluations of the objective function, it will exploit well for closed-form mathematical functions even with many local optima, such as the metamodel calibration objective functions and the estimated parameterized response surface.

8.4 APPLICATION

8.4.1 Heterogeneous Data

Both fixed sensors and probe vehicle data are collected to validate the simulation results in terms of various levels, e.g., MFDs of freeways, network-wide statistics of major arterials, and route travel time reliability. This chapter incorporates 6-month empirical observations in the 15-min time interval by loop/microwave detectors in the freeway network (from January 1 to June 30, 2013). The dataset includes lane-by-lane speed, time occupancy, and vehicle counts collected by 95 fixed detection stations along I-495, I-95, and I-270 [39]. Each station (indicated by a triangular) measures 2 through 5 lanes of the same director for every 15 min. Field measurements on weekends were excluded. Eventually, valid measurements of 130 weekdays were obtained.

The data used for the traffic flow validation of major arterials were collected via the Vehicle Probe Project Suite at the University of Maryland, College Park [39]. The primary source of the INRIX probe data includes GPS-equipped commercial fleet, delivery and taxi vehicles, mobile/smartphone probe, and driver generated report. Both historical and real-time speed data, travel time index, travel time reliability metrics, queue measurements, statewide bottleneck ranking, and corridor congestion charts by road segment for most major arterials of the network are available.

Three models of OD matrices (single occupancy vehicle, high occupancy vehicle, and trucks) are constructed for the simulator, based on demand data from the regional planning model. The dynamic OD matrices and parameters for the simulation model were calibrated using field traffic counts in [36]. Field collected intersection signal timing of the network is also included.

8.4.2 Simulation Network

An urban network of arterials and freeways were established using an open source simulation-based DTA model, DynusT [37]. All freeways and major arterials illustrated in Fig. 8.2 are included in the network, including 201 traffic

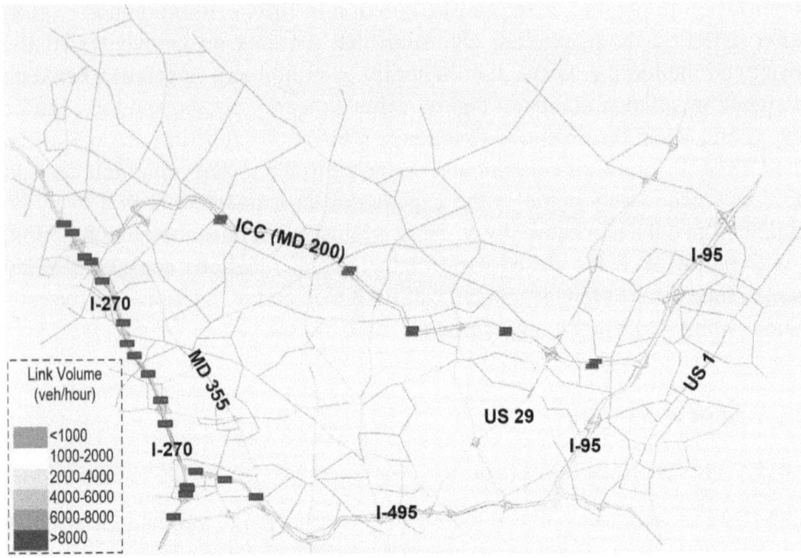

FIG. 8.2 Simulated spatial distribution of AM peak traffic flow.

analysis zones, 1077 nodes, and 2158 links. To test the effectiveness of the SBO approach presented in Section 8.3, a case study on optimizing the toll charges of the freeway network in both peak and off-peak hours is conducted. The large-scale network includes all freeways (I-495, I-95, I-270, and I-370) and major arterials (e.g., US-1, MD-355, MD-97, MD-650, and MD-28), most minor arterials, and some important local streets in central and eastern Montgomery County and northwestern Prince George's County in the State of Maryland.

Three heavily congested freeways in this study area are considered potential alternative toll routes: I-495, I-95, and I-270. The I-495 segment widely known as the Capital Beltway is a 16.5-mile freeway that, in the study area, surrounds Washington, D.C. Continuing west from the College Park Interchange, I-495 crosses into Montgomery County, Maryland, entering a heavily developed and populated area. It interchanges with MD 212, MD 650, MD 193, and US 29. Southbound US 1 is the main route into downtown College Park, continuing into Washington. I-95 is one of the most heavily traveled Interstate Highways between Washington, D.C., and Baltimore. The route utilizes the southern and eastern half of the Capital Beltway to circumnavigate Washington, D.C. and is cosigned with I-495 along that route. I-270, known as the Washington National Pike, is a 15.5-mile freeway segment in Montgomery County, State of Maryland. It travels between I-495 north of Bethesda, and MD 124 in the city of Gaithersburg, Montgomery County. The left lane on each side is used as an HOV lane in the northbound direction between 3:30 p.m. and 6:30 p.m. weekdays and in the southbound direction between 6:00 a.m. and 9:00 a.m. weekdays. In this chapter, the unit mileage toll rates of these three freeways will be optimized.

Fig. 8.2 shows a simulation snapshot of spatial traffic flow distributions at 8 a.m. during the 24-h DTA and simulation. It can be seen that the largest vehicle volumes are on three freeways I-95, I-495, and I-270. Several major arterials (e.g., MD 355) and highways (e.g., US 29) are congested, too. At the end of 20 iterations of DTA and mesoscopic simulation, the CPU time of operation is 40.03 h (CPU time of simulation is 13.36 h, CPU time of assignment is 26.66 h) using less than 5 min for one iteration of DTA with a 2.66 GHz-quad CPU and 4 GB-Ram computer.

After the 24-h demand loading, a number of 2,149,095 vehicles were generated in the study network. 2,125,592 (98.91%) vehicles completed their trip at the end of simulation. For those out-of-network vehicles, 90.43% (1,922,072) are single occupancy auto vehicles, 4.52% (96,133) are high occupancy vehicles, and 5.05% (107,387) are trucks. The average trip time is 16.917 min, the average stop time is 3.087 min, and the average trip distance is 8.086 miles.

8.4.3 Validation

Fig. 8.3 validates the simulation results by comparing the freeway network average speed, density and flow with the 6-month traffic flow data at fixed sensors. The 90% confidence intervals (CI) are estimated by the 130-weekday data. It can be seen that the simulation matches well with historical measurements considering that the simulation model was calibrated by an independent dataset, i.e., annual average daily traffic counts from 15 freeway stations and turning movement of 35 arterial roads. It is not surprising that arterials traffic dynamics are different from freeways.

By investigating all completed vehicular trajectories, Fig. 8.4A plots the average travel time rate and its S.D. for every minute. The linear relationship between σ_τ and $\bar{\tau}$ is verified. Different from other studies, the 24-h simulation results are separated into several periods in terms of their slopes in the linear relationship. It can be seen that the 1-min average travel time rates during 5–7 p.m. are larger than 2.5 min/mile, indicating the most congested period of the whole day. The corresponding S.D. values are larger than 1 min/mile, which is the largest, too. Fig. 8.4B–D show the relationships between link-based and path-based network-wide statistics.

The simulated and measured time-varying network-wide average speeds of the arterials are compared. In the INRIX datasets, 5-min speed data of weekdays over the whole year of 2012 are available for every road segment of most arterials, but road densities and vehicle counts are not known because probe vehicles consist of only a small part of all road users. The aggregate relative errors of the planning time and mean travel time of all major arterials are 7.74% and 15.24%, respectively.

Travel time distributions and variations (over the time of day) of highway segments have been widely measured and estimated in field applications, e.g., loop data [40, 41]. A series of statistical reliability metrics for a specific link or

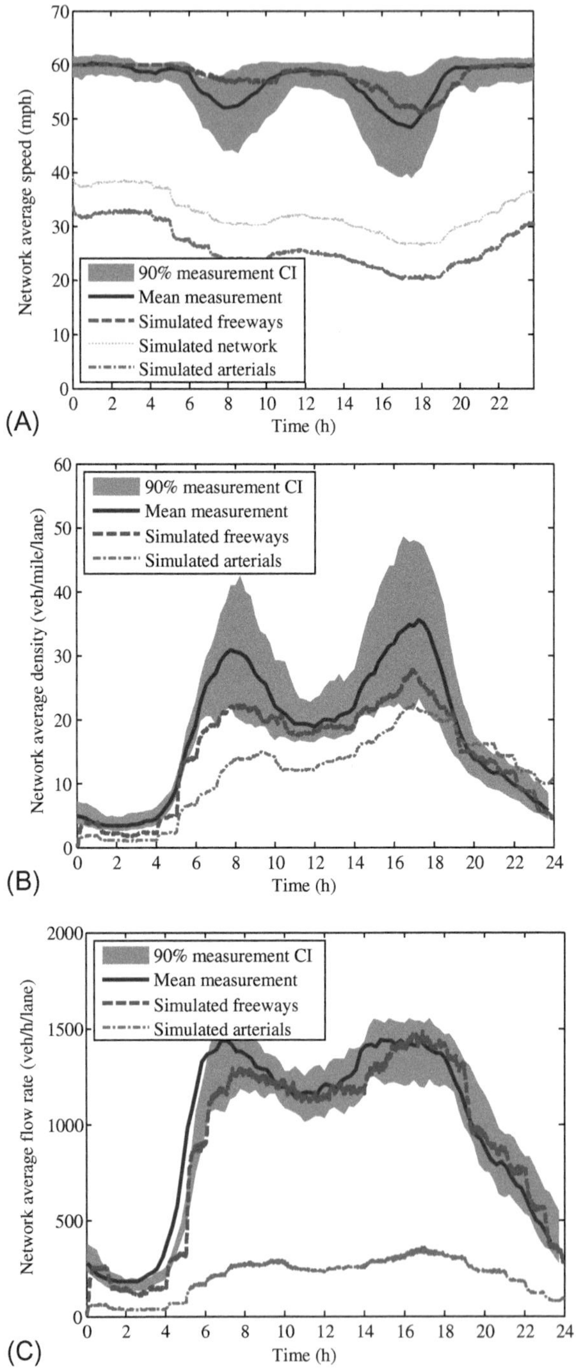

FIG. 8.3 Comparisons of the simulated and measured freeway traffic flow. (A) V_t^{freeway}. (B) K_t^{freeway}. (C) Q_t^{freeway}.

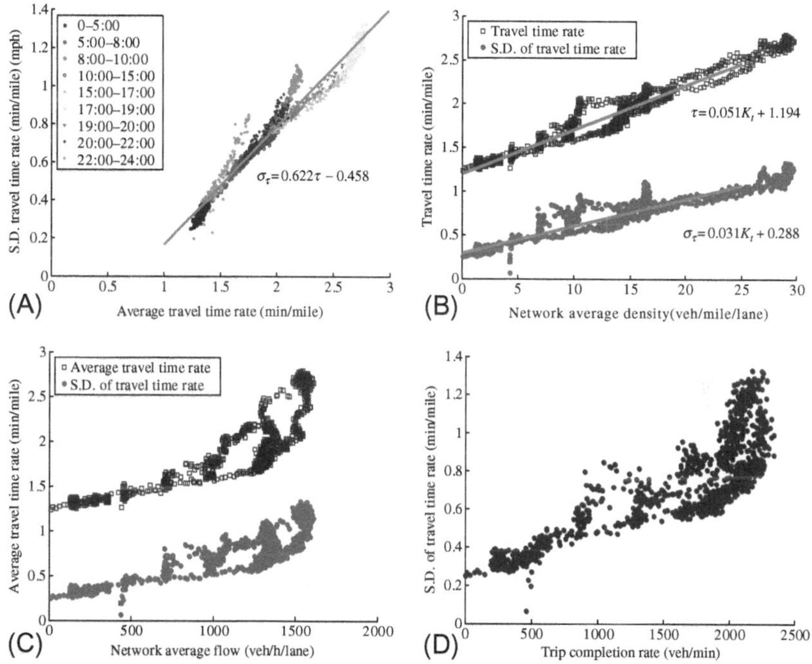

FIG. 8.4 Simulated relationships between link-based and path-based network-wide statistics. (A) $\bar{\tau}_t$ vs. σ_τ. (B) K_t vs. $\bar{\tau}_t$ and σ_τ. (C) Q_t vs. $\bar{\tau}_t$. (D) Trip completion rate vs. σ_τ.

corridor were recommended by [42], e.g., the percentile of travel time index, buffer index, and planning time index. Measuring path travel time reliability requires substantial numbers of observations across days, for any given time of day. It suggests that the most common measure of unreliability, the standard deviation of travel time across days, will only imperfectly capture traveler preferences [43]. For this reason, the study network relies instead on measures of the upper tail of the distribution of historical travel times, e.g., the 95th and 50th percentile travel times. Measured planning time means how much total time a traveler should allow to ensure on-time arrival (the 95th percentile travel time). It includes typical delay as well as unexpected delay, indicating the near-worst case travel time. This method is perhaps the simplest method to measures travel time reliability. It estimates how bad delay will be on specific routes during the heaviest traffic days. Measured travel time is the average time it will take to drive along the stretch of road (distance traveled/speed). The gap between the planning time and the average travel time is called buffer time, which could be explained as the extra time (or time cushion) that travelers must add to their average travel time when planning trips to ensure on-time arrival due to travel time variability.

Table 8.1 shows the measured and simulated planning time and mean travel time on a number of major arterial routes. The 5-min time interval

TABLE 8.1 Route-by-Route Validation With Probe Vehicle Travel Time Statistics

Major Arterials (Length)	Directions	Sample Trajectories	Measured Planning Time (min)	Measured Travel Time (min)	Simulated Planning Time (min)	Simulated Travel Time (min)	Planning Time Relative Error (%)	Travel Time Relative Error (%)
US-1	NB	513	10.94	8.15	10.14	8.81	7.35	8.10
(5.10 mile)	SB	1271	10.54	7.99	10.60	9.33	0.57	16.81
MD-97	NB	171	29.62	21.52	25.36	22.08	14.39	2.62
(12.16 mile)	SB	107	27.87	20.56	42.01	25.87	50.74	25.81
MD-650	NB	142	20.28	16.45	25.85	21.20	27.49	28.90
(10.38 mile)	SB	64	20.18	16.33	44.26	22.80	119.34	39.63
MD-586	EB	201	15.87	11.84	14.08	11.27	11.29	4.80
(6.23 mile)	WB	96	14.64	11.51	13.07	11.60	10.74	0.82
MD-355	NB	92	13.12	9.64	12.65	9.64	3.60	0.04
(4.82 mile)	SB	346	13.73	9.65	11.26	9.25	17.99	4.19
MD-28	EB	292	15.77	11.9	16.61	14.54	5.35	22.19
(7.39 mile)	WB	281	15.02	11.63	16.73	14.49	11.38	24.59

MD-193	EB	45	12.11	9.7	15.19	13.82	25.45	42.47
(4.98 mile)	WB	4	12.71	9.62	14.01	13.40	10.23	39.32
MD-185	NB	1853	12.89	10.38	13.79	11.64	6.98	12.10
(5.95 mile)	SB	2196	13.67	9.89	13.85	11.65	1.29	17.81
MD-182	NB	50	12.19	10.62	14.49	13.11	18.87	23.43
(6.54 mile)	SB	256	12.56	10.56	15.40	13.29	22.59	25.89
MD-115	EB	107	12.77	10.15	14.22	12.20	11.32	20.16
(5.87 mile)	WB	141	12.63	10.12	14.42	12.93	14.19	27.81
MD-108	EB	85	23.52	20.18	23.01	21.00	2.19	4.08
(13.15 mile)	WB	166	22.43	19.93	22.45	20.05	0.08	0.61
All major arterials	N/A[a]	8479	N/A	N/A	N/A	N/A	7.74[b]	15.24[b]

[a]*N/A: not applicable.*
[b]*Weighted average relative error across all routes, using the number of samples as weights.*

measurements of 261-weekday probe vehicle data in the whole year of 2012 enable us to estimate statistical quantities based on the travel time distribution. Most reliability measures are calculated from the day-to-day distribution of travel times on a particular route, that is, the travel time variability that vehicles experience on a particular time-of-day and day-of-week period over a longer time period. Fig. 8.5 shows the detailed comparisons. Each circle point represents a vehicle trip that passes through the corresponding route. If the vehicular trajectory covers the specific route, the travelers' entering time and leaving time indices for each route are used to estimate its experienced travel time. The path travel time reliability difference of simulations can be partially explained as the close intertwining with the signal timing along major arterials.

8.4.4 SBO Results

In this section, we study journeys during the extended peak (4–11 a.m.) including the morning peak (6–9 a.m.), two-hour off-peak periods (5–6 a.m. and 9–10 a.m.), one-hour warm-up simulation (4–5 a.m.), and one-hour discharge simulation (10–11 a.m.). In the following, all traffic flow statistics are calculated for the middle five-hour period, i.e., 5–10 a.m. A key purpose of time-varying tolls is to manage congestion. Raising the toll rate during periods of high traffic volumes helps reduce the number of vehicles using the roadway, keeps the highway relatively congestion-free, and provides travelers with reliable travel times. The value of time is US$ 15 per hour, estimated from a household survey [1]. For each iteration and each OD-departure time combination, the number of vehicles is dependent on the relative gap function value, and vehicles with longer travel time are prioritized for path updating. Fixed commuting demands are assumed in extended morning peak hours for the aforementioned road network in the State of Maryland. In this way, the specific optimization problem for the study network is given by:

$$\min_{\mathbf{x}\in\mathbb{R}^5} \mathrm{E}[f(\mathbf{x})] = \mathrm{E}[f(x_1, x_2, x_3, x_4, x_5)] \tag{8.17}$$

$$\text{s.t.} \quad \mathbf{x}_{\min} \le \mathbf{x} \le \mathbf{x}_{\max} \tag{8.18}$$

where $f(\mathbf{x})$ represents the unknown true objective function of the whole network given the input $\mathbf{x}$, which is a five-dimensional decision variable vector (i.e., I-495 EB peak distance-based toll in US$/mile, I-495 WB peak distance-based toll in US$/mile, I-95 SB peak distance-based toll in US$/mile, I-270 SB peak distance-based toll in US$/mile, and off-peak/peak distance toll ratio in percentage). The scenarios considered are of the average trip travel time minimization, throughput maximization, toll revenue maximization, and desirability maximization. The box constraints are $\mathbf{x}_{\min} = [0,0,0,0,0]^{\mathrm{T}}$ and $\mathbf{x}_{\max} = [0.3, 0.3, 0.3, 0.3, 100\%]^{\mathrm{T}}$, which are lower and upper boundaries for freeway unit mileage toll rates, respectively.

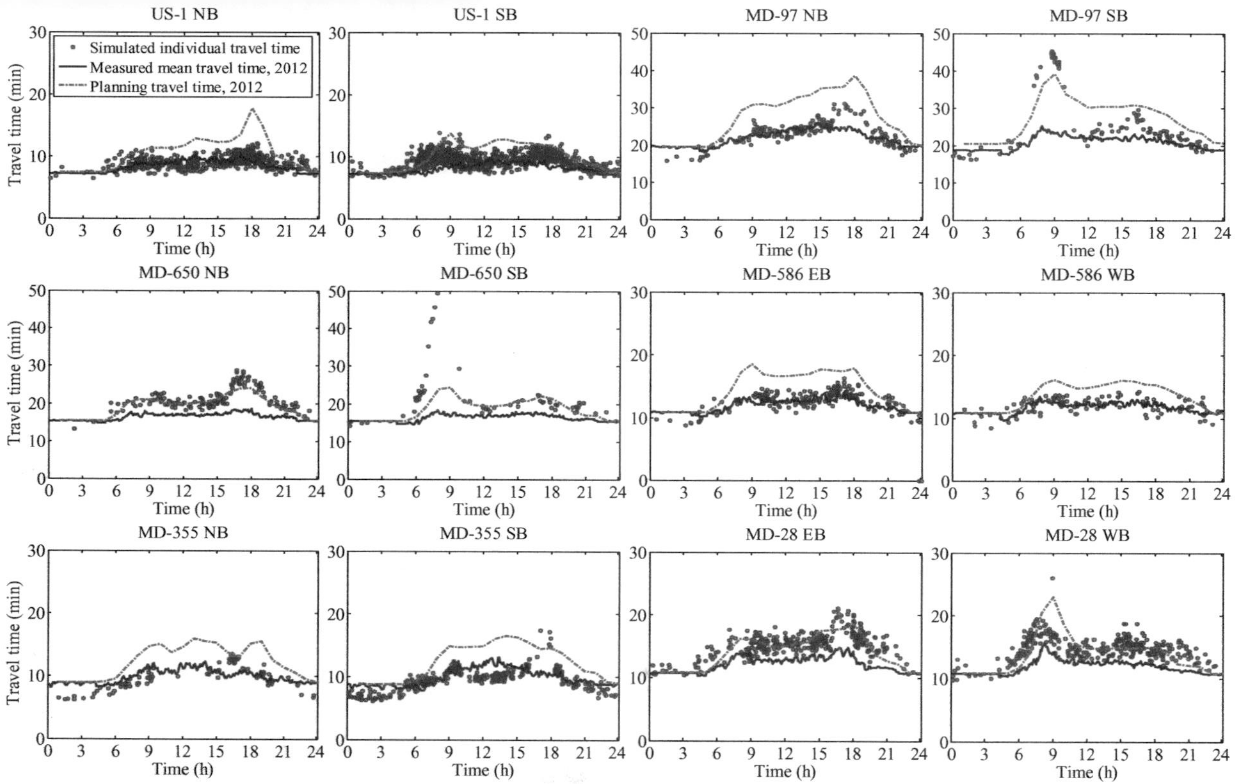

FIG. 8.5 Comparison of simulated trip travel time with historical INRIX route travel time.

(Continued)

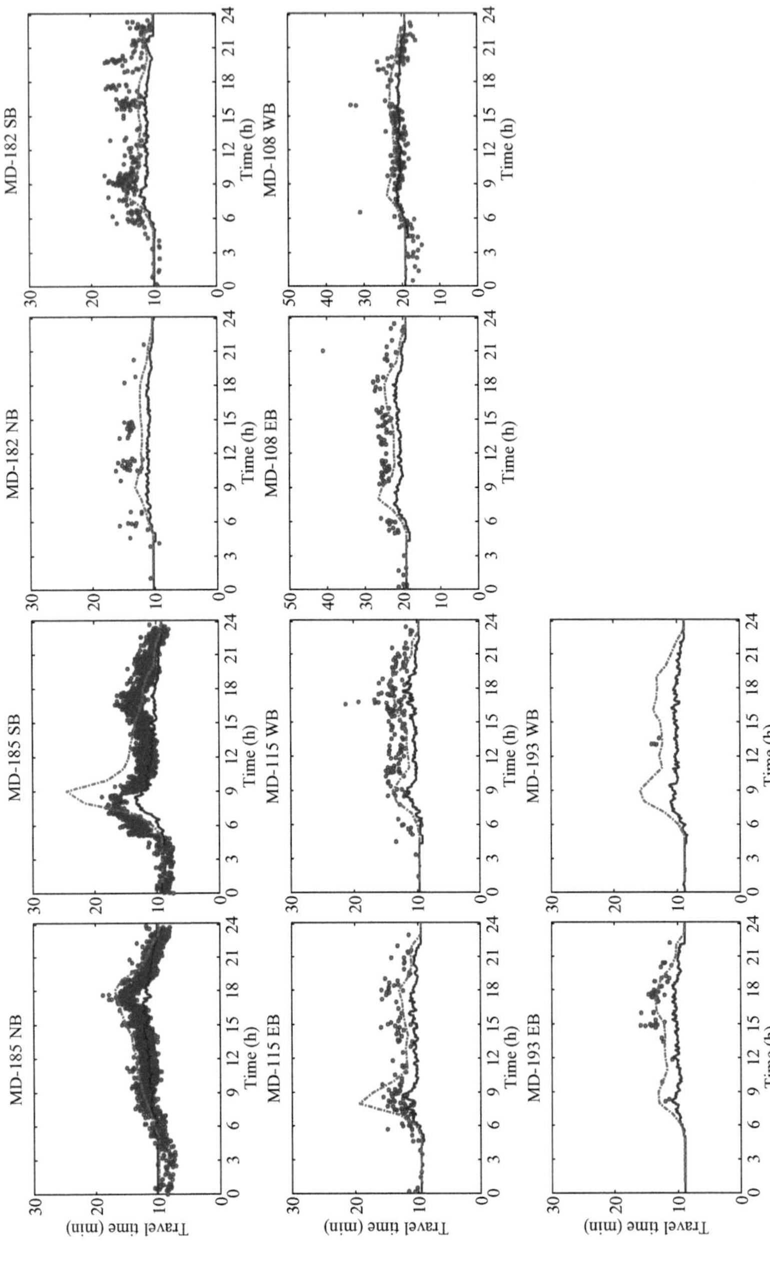

FIGURE 8.5—cont'd

The Kriging surrogate model is identified to approximate the real response surface in the simulation and is applied to search for the optimal mileage based toll rates in the aforementioned four types of objectives.

Before processing the SBO approach, the optimized sample plan, i.e., **X**, is obtained by LHS, including 100 design points plus upper and lower bounds (the minimum toll $\mathbf{x}_{min} = \mathbf{0}$ and the maximum toll $\mathbf{x}_{max}$). The baseline is the lower bound scenario, which has no freeway toll but with HOV restrictions in peak hours. DynusT obtains valid results when the convergence is achieved after several times of assignments and vehicular platoon simulations. For each $\mathbf{x} \in \mathbf{X}$, 10 iterations of DTA and traffic simulations are run to achieve convergence, which generally takes 7.2 h on average to finish each input sample. So the total computational time spent is about 734 h. DynusT obtains valid results when the convergence is achieved after 10 iterations of DTA and vehicular platoon simulations. The relative gaps were found to be below 3% for all experiments. The Kriging model helps reduce tremendous computation time compared to a traditional optimization that needs to evaluate a large number of possible solutions through the entire feasible domain.

Fig. 8.6A–C) shows the simulation outputs of 102 initial DoE inputs in terms of average trip travel time, network throughput, and toll revenue. To enhance

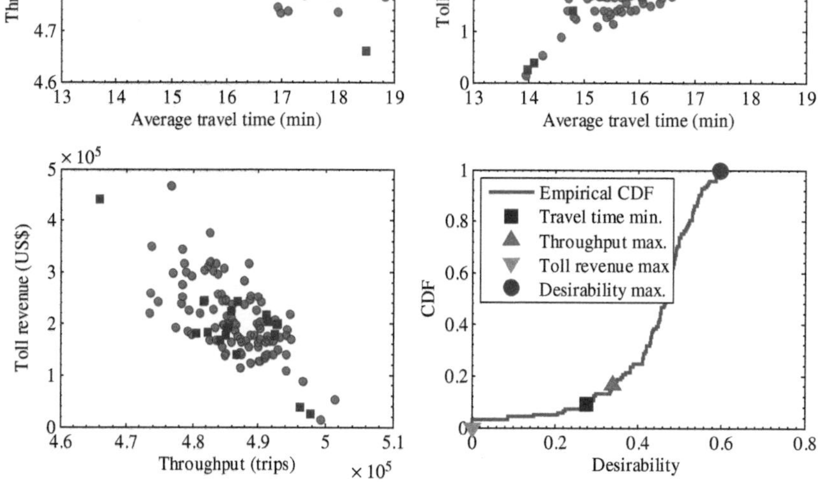

FIG. 8.6 Individual objective functions and empirical cumulative distribution of desirability.

the fine tuning of the global optima for the overall desirability function, two-stage procedures of initial DoE (the first stage) and infill points based on the expected improvement maximization criterion are incorporated. The second stage conducts the exploitation process using estimated standard deviation information to select an infill sample with the maximum expected improvement. Constrained by the computational costs, 20 infill points are explored and shown by the squares in Fig. 8.6A–C). The empirical cumulative distribution function (CDF) of the overall desirability of initial and infill points is shown in Fig. 8.6D. The desirability maximization scenario outperforms other three scenarios in the overall desirability obviously.

The optima are estimated by the results of the Kriging model using 122 points, including the initial samples and 20 infill points with the expected improvement maximization criterion [1]. The travel time minimization scenario obtains the optimal solution of the lower bound, i.e., $\hat{\mathbf{x}}^* = \mathbf{x}_{\min}$, which is in conformity with the intuitive judgment that when all freeways open to the public, then the network travel time would reach the minimum. The throughput maximization scenario is analogous to the former scenario, while the toll revenue maximization scenario is achieved when the upper tolls are implemented, i.e., $\hat{\mathbf{x}}^* = \mathbf{x}_{\max}$, which indicates a large number of travelers prefer to choose to use freeways even under the upper toll rates.

The simulation generates statistics of the network performance every minute. As link speed fluctuates significantly at the one-minute interval, Fig. 8.7 shows 5-min moving average speeds of major arterials during the extended a.m. peak under four scenarios. The average speed difference between the baseline toll and the maximum throughput case is tiny because both the

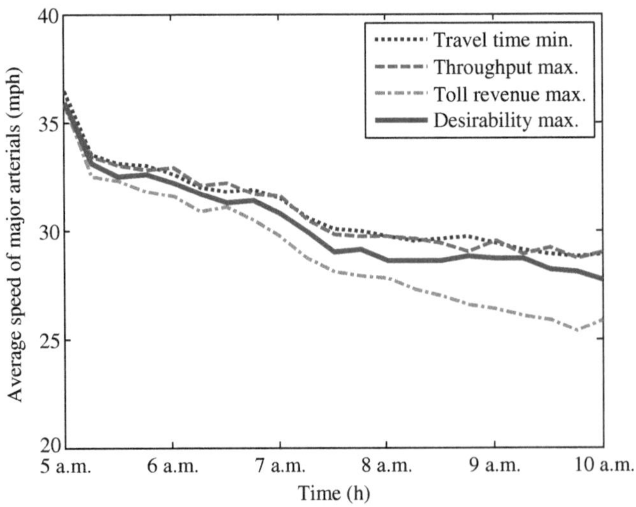

FIG. 8.7 Comparison of major arterial average speeds of multiple objective functions.

average travel time objective and the throughput are consistent with each other. However, the maximum toll revenue scenario drives a certain amount of traffic from freeways to major arterials, resulting in a smaller average speed than the former two scenarios. While the total desirability maximization scenario exhibits a medium congestion level. Various network-wide performance measures are also compared among the four toll scenarios. Fig. 8.8A shows the network-wide average trip travel time along with the departure time. Fig. 8.8B compares time-varying vehicle throughput for the four scenarios. While the simulations of travel time minimization, throughput maximization, and total desirability maximization perform similar throughput patterns, we zoom in on the last statistical hour, i.e., 9–10 a.m., in Fig. 8.8B. Fig. 8.8C shows the time-varying toll revenue. In general, the total desirability maximization scenario is a moderate mileage based toll strategy compared with the zero or near zero toll and the maximum toll scenarios because all three individual produced objectives are not bad, and thus generate the highest desirability.

8.5 CONCLUSIONS

This chapter develops a SBO framework by integrating metamodels with mesoscopic simulation-based dynamic traffic assignment models for large-scale transportation optimization problems. The performance of the proposed SBO framework is evaluated by addressing the congestion pricing of freeways in a real-world large-scale network that consists of arterials and freeways. The simulation-based dynamic traffic assignment model is validated by a representation of MFDs using locally collected high-quality data such as fixed traffic flow detections and probe vehicle travel time measurements.

The SBO framework is effectively applied and tested in a real-world transportation network using DynusT. Although the computational burden is heavy and usually not affordable when using a simulator to evaluate expensive-to-evaluate objective functions with high dimensional decision variables, the SBO approach is capable of solving such problems without closed-form objective functions. The main benefit in computational time savings can be achieved by using metamodels to construct response surfaces for predicting optimal toll charges. The SBO framework established in this chapter can support joint optimization of transportation policies, planning applications, and operational strategies in future research.

ACKNOWLEDGMENTS

This research is financially supported by Zhejiang Provincial Natural Science Foundation of China [LR17E080002], National Natural Science Foundation of China [51508505, 71771198], Key Research and Development Program of Zhejiang [2018C01007], Fundamental Research Funds for the Central Universities [2017QNA4025]. The author is highly thankful to Prof. Lei Zhang, Dr. Chenfeng Xiong, and Dr. Zheng Zhu, University of Maryland,

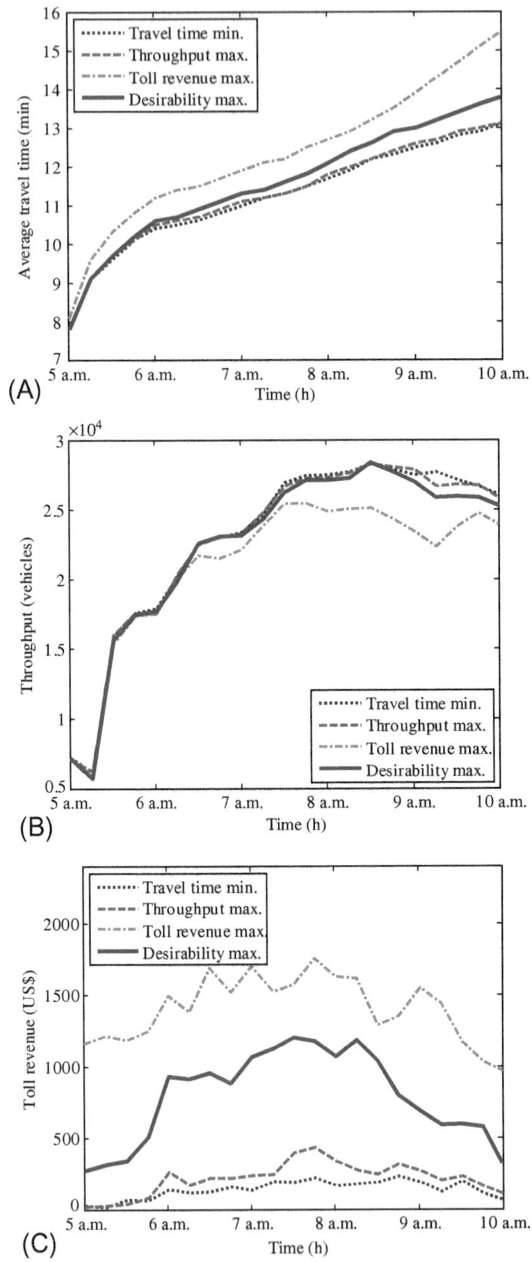

FIG. 8.8 Comparison of multiple objective functions. (A) Network-wide average trip travel time. (B) Vehicle throughput. (C) Toll revenue.

College Park, MD, USA, and Dr. Xiang He, China Academy of Civil Aviation Science and Technology, International Aviation Division, Institute of Air Transport, for their valuable discussions, constructive comments, and technical support.

REFERENCES

[1] X. Chen, L. Zhang, X. He, C. Xiong, Z. Li, Surrogate-based optimization of expensive-to-evaluate objective for optimal highway toll charges in transportation network, Comput-Aided Civ. Inf. Eng. 29 (5) (2014) 359–381.

[2] X. Chen, Z. Zhu, L. Zhang, in: Simulation-based optimization of mixed road pricing policies in a large real-world network, European Transport Conference, Frankfurt, Germany, September 29–October 1, 2014.

[3] X. Chen, Z. Zhu, L. Zhang, Simulation-based optimization of mixed road pricing policies in a large real-world network, Transp. Res. Procedia 8 (2015) 215–226.

[4] X. Chen, Z. Zhu, X. He, L. Zhang, Surrogate-based optimization for solving mixed integer network design problem, J. Transp. Res. Board: Transp. Res. Rec. 2497 (2015) 124–136.

[5] X. Chen, L. Zhang, X. He, C. Xiong, in: Simulation-based pricing optimization for improving network-wide travel time reliability, The 6th International Symposium on Transportation Network Reliability, Nara, Japan, August 2–3, 2015.

[6] X. Chen, C. Xiong, X. He, Z. Zhu, L. Zhang, Time-of-day vehicle mileage fees for congestion mitigation and revenue generation: a simulation-based optimization method and its real-world application, Transp. Res. C Emerg. Technol. 63 (2016) 71–95.

[7] X. He, X. Chen, C. Xiong, Z. Zhu, L. Zhang, Optimal time-varying pricing for toll roads under multiple objectives: a simulation-based optimization approach, Transp. Sci. 51 (2) (2017) 412–426.

[8] C. Osorio, M. Bierlaire, A simulation-based optimization framework for urban transportation problems, Oper. Res. 61 (6) (2013) 1333–1345.

[9] C. Osorio, L. Chong, A computationally efficient simulation-based optimization algorithm for large-scale urban transportation problems, Transp. Sci. 49 (3) (2015) 623–636.

[10] C. Osorio, K. Nanduri, Energy-efficient urban traffic management: a microscopic simulation-based approach, Transp. Sci. 49 (3) (2015) 637–651.

[11] C. Osorio, K. Selvam, Simulation-based optimization: achieving computational efficiency through the use of multiple simulators, Transp. Sci. 51 (2) (2017) 395–411.

[12] C. Zhang, C. Osorio, G. Flötteröd, Efficient calibration techniques for large-scale traffic simulators, Transp. Res. B Methodol. 97 (2017) 214–239.

[13] M. Song, M. Yin, X. Chen, L. Zhang, M. Li, in: A simulation-based approach for sustainable transportation systems evaluation and optimization: theory, systematic framework and applications, The 13th COTA International Conference of Transportation Professionals, Shenzhen, China, August 13–16, 2013.

[14] L. Zhang, X. Chen, X. He, C. Xiong, in: Bayesian stochastic kriging metamodel for active traffic management of corridors, IIE Annual Conference and Expo, Montreal, Canada, May 31–June 3, 2014.

[15] X. Chen, M. Yin, M. Song, L. Zhang, M. Li, Social welfare maximization of multimodal transportation: theory, metamodel, and application to Tianjin Ecocity, China, J. Transp. Res. Board: Transp. Res. Rec. 2451 (2014) 36–49.

[16] X. Chen, X. He, C. Xiong, L. Zhang, in: A Bayesian stochastic kriging metamodel for simultaneous optimization of travel behavioral responses and traffic management, The 94th Annual Meeting of Transportation Research Board, Washington, DC, January 11–15, 2015.

[17] X. He, X. Chen, C. Xiong, Z. Zhu, L. Zhang, in: Integrated optimization of transportation demand management and traffic operations using simulation: a bootstrapped support vector regression method considering the statistical distribution of simulation noise, The 94th Annual Meeting of Transportation Research Board, Washington DC, January 11–15, 2015.
[18] Z. Zhu, C. Xiong, X. Chen, X. He, L. Zhang, Agent-based microsimulation approach for design and evaluation of flexible work schedules, J. Transp. Res. Board: Transp. Res. Rec. 2537 (2015) 167–176.
[19] C. Xiong, X. Chen, X. He, L. Zhang, in: Simulation-based optimal travel information provision strategies: an agent-based approach under uncertainty, The 6th International Symposium on Transportation Network Reliability, Nara, Japan, August 2–3, 2015.
[20] Z. Zhu, C. Xiong, X. Chen, L. Zhang, in: Calibrating supply parameters of large-scale dynamic traffic assignment models with simulation-based optimization, The 96th Annual Meeting of Transportation Research Board, Washington, DC, January 8–11, 2017.
[21] X. Chen, S. Zhang, M. Zahiri, in: Cellular signaling data driven simulation-based dynamic traffic assignment and its applications to a real-world road network, The 19th International IEEE Conference on Intelligent Transportation Systems, Rio de Janeiro, Brazil, November 1–4, 2016.
[22] M. Li, X. Lin, X. Chen, A surrogate-based optimization algorithm for network design problems, Front. Inf. Technol. Electron. Eng. 18 (11) (2017) 1693–1704.
[23] T.J. Santner, B.J. Williams, W.I. Notz, The design and analysis of computer experiments, Springer, New York, 2003.
[24] D. Huang, T.T. Allen, W.I. Notz, N. Zeng, Global optimization of stochastic black-box systems via sequential Kriging meta-models, J. Glob. Optim. 34 (3) (2006) 441–466.
[25] C.E. Rasmussen, C.K.I. Williams, Gaussian Processes for Machine Learning, MIT Press, Cambridge, 2006.
[26] R. Barton, M. Meckesheimer, Metamodel-Based Simulation Optimization. Handbooks in Operations Research and Management Science, Elsevier B.V, Amsterdam, 2006.
[27] D.C. Montgomery, Design and Analysis of Experiments, Wiley, 2008.
[28] A. Forrester, A. Sobester, A. Keane, Engineering Design via Surrogate Modelling: A Practical Guide, Wiley, 2008.
[29] R.J. Smeed, Road capacity of city centers, Traffic Eng. Control 8 (7) (1966) 455–458.
[30] H. Mahmassani, J.C. Williams, R. Herman, Investigation of network-level traffic flow relationships: some simulation results, Transp. Res. Rec.: J. Transp. Res. Board 971 (1984) 121–130.
[31] N. Geroliminis, C.F. Daganzo, Existence of urban-scale macroscopic fundamental diagrams: some experimental findings, Transp. Res. B 42 (9) (2007) 759–770.
[32] C.F. Daganzo, Urban gridlock: macroscopic modeling and mitigation approaches, Transp. Res. B 41 (1) (2007) 49–62.
[33] C.F. Daganzo, N. Geroliminis, An analytical approximation for macroscopic fundamental diagram of urban traffic, Transp. Res. B 42 (9) (2007) 771–781.
[34] N. Geroliminis, J. Sun, Properties of a well-defined macroscopic fundamental diagram for urban traffic, Transp. Res. B 45 (3) (2011) 605–617.
[35] C. Xiong, X. Chen, X. He, X. Lin, L. Zhang, Agent-based en-route diversion simulation: dynamic behavioral responses and network performance represented by macroscopic fundamental diagram, Transp. Res. C: Emerg. Technol. 64 (2016) 148–163.
[36] C. Xiong, Z. Zhu, X. He, X. Chen, S. Zhu, S. Mahapatra, G.-L. Chang, L. Zhang, Developing a 24-hour large-scale microscopic traffic simulation model for the before-and-after study of a new tolled freeway in the Washington DC-Baltimore region, ASCE J. Transp. Eng. 141 (6) (2015).

[37] Y.-C. Chiu, L. Zhou, H. Song, Development and calibration of the anisotropic mesoscopic simulation model for uninterrupted flow facilities, Transp. Res. B Methodol. 44 (1) (2010) 152–174.

[38] D.R. Jones, C.D. Perttunen, B.E. Stuckman, Lipschitzian optimization without the Lipschitz constant, J. Optim. Theory Appl. 79 (1) (1993) 157–181.

[39] CATT Lab, The Center for Advanced Transportation Technology Laboratory, http://www.cattlab.umd.edu/, 2013.

[40] C. Chen, A. Skabardonis, P. Varaiya, Travel-time reliability as a measure of service, Transp. Res. Rec.: J. Transp. Res. Board 1855 (2003) 74–79.

[41] K. Lyman, R.L. Bertini, Using travel time reliability measures to improve regional transportation planning and operations, Transp. Res. Rec.: J. Transp. Res. Board 2046 (2008) 1–10.

[42] R. Margiotta, T. Lomax, M. Hallenbeck, R. Dowling, A. Skabardonis, S. Turner, Analytical Procedures for Determining the Impacts of Reliability Mitigation Strategies, SHRP 2 Report S2-L03-RR-1, 2013.

[43] D. Brownstone, K.A. Small, Valuing time and reliability: assessing the evidence from road pricing demonstrations, Transp. Res. A 39 (4) (2005) 279–293.

Chapter 9

Network Modelling and Resilience Analysis of Air Transportation: A Data-Driven, Open-Source Approach

Xiaoqian Sun and Sebastian Wandelt

National Key Laboratory of CNS/ATM, School of Electronic and Information Engineering, Beihang University, Beijing, People's Republic of China

Chapter Outline

9.1 INTRODUCTION

Air transportation is a critical network infrastructure for a nation and it is becoming extremely important for public policy considerations. Disruption of air transportation systems, either due to extreme weather conditions or terrorist attacks, can lead to huge economic losses. The eruption of the Eyjafjallajoekull volcano in 2010 caused airlines losses of approximately 1.7 billion US dollars and more than 10 million passengers were affected [1]. An overnight snowstorm on March 12, 2013 disrupted air transport across northwestern Europe, in particular Frankfurt airport was closed and airlines canceled about 700 flights. In order to avoid such high socio-economic costs, it is critical to assess the effectiveness and robustness of air transportation systems against natural or intentional disruptions.

Many studies have been performed on air transportation networks, we only summarize the general trends here with links to the relevant literature.

Data-Driven Solutions to Transportation Problems. https://doi.org/10.1016/B978-0-12-817026-7.00010-2

The analysis of air transportation systems as complex networks has a history of more than ten years [2–4]. Many of these studies focus on the importance of nodes as well as the role of nodes in the networks [5,6]. Several studies have analyzed the commonalities and dissimilarities of sub networks [7,8]. Another facet, also based on complex network techniques, is the estimation of resilience of air transportation systems for the global network and its subnetworks [9–14]. Another line of research investigates the network at different levels of fractality [15–17] and aggregations of airports into larger systems, so called Multiple Airport Regions [18–22]. Other studies analyze the impact of delay and other events on the efficiency and effectiveness of the network [23–25].

Overall, many studies have been published on the evaluation of air transportation networks. The major limitations of these studies are summarized as follows:

(1) Use of proprietary data: Studies often make use of either commercial or hand-collected data. The unavailability of public data makes research on air transportation networks unnecessarily difficult and not reproducible. The usage of public air transportation data leverages these effects, by increasing the accessibility and impact of such studies.

(2) Nonstandard data preparation: Existing studies have not established a commonly agreed method of data preparation and analysis. Essentially, each study has a slightly different design, making it difficult to cross-compare studies.

(3) Closed-source analysis: In other research disciplines, for instance, Bioinformatics, it has become standard to publish the source codes of analysis. This is not yet the case in air transportation research. Releasing the source code makes research reproducible and increases its overall impact.

(4) Limited facets in each study: Many studies investigate a few facets of air transportation only. Merging the results for an overall picture is difficult, given the concerns on methodology and data usage discussed above.

In this chapter, we aim to provide a data-driven, open-source framework for network modeling as well as resilience analysis of air transportation systems. We design, implement, and disseminate such a framework for analysis. The framework is implemented in Python and made freely available for academic use.

The rest of the chapter is organized as follows. Section 9.2 introduces the dataset we used in this study. The network modeling of air transportation is presented in Section 9.3. Several properties of the worldwide air transportation network are analyzed in Section 9.4. The chapter is concluded with Section 9.5.

9.2 DATA PREPARATION

First, this section introduces the dataset from http://openflights.org which we used for our experiments. The dataset represents a snapshot of the global air transportation network in 2017. We describe the setup and present a reproducible workflow for data preparation.

While there are many datasets available for air transportation systems, most of them are commercial and require a substantial amount of money to purchase. Money as a kind of gatekeeper makes it harder for researchers to perform studies and air transportation research less accessible to the general public. The website http://openflights.org is a donation-funded service which provides current air transportation routes as extracted from Airline Route Mapper and other data sources. The databases are being made available under the Open Database License, which means that anybody can use the data as long as the data source and data license are provided and any derived works are made available to the public with a free license. This makes the dataset particularly attractive for air transportation research. The data can be downloaded from https://openflights.org/data.html. In the following, we describe the structure and semantics of the supplied datasets, as far as relevant to our study.

The website openflights.org distinguishes three types of datasets: Airports, Airlines, and Routes. All datasets are represented following a relational data model, where columns present attributes of an entity and columns list instances of that entity. The airport entity describes a set of more than 7000 airports worldwide, with their name, location, altitude, and more information. We present an extract of the airport relation in Table 9.1. Besides their IATA and ICAO code, each airport has a unique numerical identifier which is used by openflights.com as a primary key for linking to other entities.

The airline entity collects information on around 6000 airlines worldwide. An extract of that relation is presented in Table 9.2. Similar to airports, airlines have an internal numerical identifier as a primary key, which is used for linking below.

The third relation, routes, describes the routes of airlines between airports, i.e., which airports are connected by which airline. An extract of relation for routes is presented in Table 9.3. The connection between the three entities, Airports, Airlines, and Routes, is induced by foreign key relationships from Routes to Airports and Airlines. For each pair of airports, we can see which airlines are serving this pair. This information is used below to derive a complex network representation of global air transportation.

9.3 AIR TRANSPORTATION NETWORK MODELING

Naturally, air transportation networks can be modeled as a weighted network, where airports are nodes, and direct flights between airports are links. The set of nodes comes from the Airport data table at openflights.org, while the set of flights are derived from the Routes table. It should be noted that we disregard information about airlines at this stage, but take it into account later for more detailed analysis. From the two tables we derived the global air transportation network; the results are visualized in Fig. 9.1. We can see that the airports are distributed over all continents, with a slight focus towards airports in North America and Western Europe. In total, we have 3246 active airports and

TABLE 9.1 An Example of Airport Entity Provided by Openflights

ID	Name	IATA	ICAO	Lat	Lon	Alt
1	"Goroka Airport"	"GKA"	"AYGA"	−6.0816898346	145.391998291	5282
2	"Madang Airport"	"MAG"	"AYMD"	−5.2070798874	145.789001465	20
3	"Mount Hagen Kagamuga …"	"HGU"	"AYMH"	−5.826789856	144.296005249	5388
4	"Nadzab Airport"	"LAE"	"AYNZ"	−6.569803	146.725977	239
5	"Port Moresby Jacksons …t"	"POM"	"AYPY"	−9.4433803558	147.2200012207	146
6	"Wewak International Airport"	"WWK"	"AYWK"	−3.5838301182	143.669006348	19
7	"Narsarsuaq Airport"	"UAK"	"BGBW"	61.1604995728	−45.4259986877	112
8	"Godthaab/Nuuk Airport"	"GOH"	"BGGH"	64.19090271	−51.6781005859	283
9	"Kangerlussuaq Airport"	"SFJ"	"BGSF"	67.0122218992	−50.7116031647	165
10	"Thule Air Base"	"THU"	"BGTL"	76.5311965942	−68.7032012939	251
…	…	…	…	…	…	…
12,057	"Shenyang Dongta Airport"	"N"	"ZYYY"	41.7844	123.4960	0

For each airport, its unique numerical identity (ID), name, IATA code, ICAO code, latitude, longitude, and altitude are included. The table is provided as a comma-separated value file. The column ID is the unique primary key for all instances. The value "N" is used for unknown/unassigned values.
Data source: http://openflights.org/data.

TABLE 9.2 An Example of Airline Entity Provided by Openflights

ID	Name	Alias	IATA	ICAO	Callsign	Country	Active
1	"Private flight"	"N"	"-"	"N/A"	"-"	"-"	"Yes"
2	"135 Airways"	"N"	"-"	"GNL"	"GENERAL"	"United States"	"No"
3	"1Time Airline"	"N"	"1 T"	"RNX"	"NEXTIME"	"South Africa"	"Yes"
4	"2 Sqn No 1 E..."	"N"	"-"	"WYT"	"-"	"United Kingdom"	"No"
5	"213 Flight Unit"	"N"	"-"	"TFU"	"-"	"Russia"	"No"
6	"223 Flight ..."	"N"	"-"	"CHD"	"CHKALOVSK-AVIA"	"Russia"	"No"
7	"224th Flight Unit"	"N"	"-"	"TTF"	"CARGO UNIT"	"Russia"	"No"
8	"247 Jet Ltd"	"N"	"-"	"TWF"	"CLOUD RUNNER"	"United Kingdom"	"No"
9	"3D Aviation"	"N"	"-"	"SEC"	"SECUREX"	"United States"	"No"
…	…	…	…	…	…	…	…
21,317	"Svyaz Rossiya"	…	"7R"	"SJM"	"RussianConnecty"	"Russia"	"Yes"

For each airline, its unique numerical identity (ID), name, alias, IATA code, ICAO code, callsign, which country it belongs to, and whether it is active or not are included. The table is provided as a comma-separated value file. The column ID is the unique primary key for all instances. The value "N" is used for unknown/unassigned values. Data source: http://openflights.org/data

TABLE 9.3 An Example of Routes Entity Provided by Openflights

Airline	Airline ID	Source	Source ID	Destination	Destination ID	Codeshare	Stops	Equip
2B	410	AER	2965	KZN	2990		0	CR2
2B	410	ASF	2966	KZN	2990		0	CR2
2B	410	ASF	2966	MRV	2962		0	CR2
2B	410	CEK	2968	KZN	2990		0	CR2
2B	410	CEK	2968	OVB	4078		0	CR2
2B	410	DME	4029	KZN	2990		0	CR2
2B	410	DME	4029	NBC	6969		0	CR2
2B	410	DME	4029	TGK	N		0	CR2
2B	410	DME	4029	UUA	6160		0	CR2
2B	410	EGO	6156	KGD	2952		0	CR2
…	…	…	…	…	…	…	…	…
ZM	19,016	OSS	2913	FRU	2912		0	734

The table is provided as a comma-separated value file. The table has no primary keys, but foreign keys into Airlines and Airports. The value "N" is used for unknown/unassigned values.
Data source: http://openflights.org/data

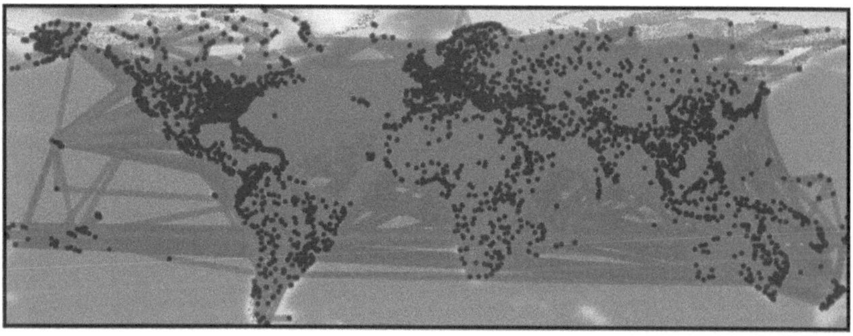

FIG. 9.1 Global air transportation network from openflights. *Notes:* Airports are visualized as *dots* and direct flight connections with *links*. In total, we have 3246 airports and 18,890 connections. Please note that all flights are visualized through the center of the figure; actual routes might be different.

18,890 flight connections. The visualization in Fig. 9.1 is based on the spatial information of nodes (latitude and longitude). However, complex networks, as a level of abstraction, often investigate networks without spatial information. In this case, we can use automatic layouts, which determine the node positions based on the network's topology.

In Fig. 9.2, we show the network as derived with the layout algorithm Fruchterman Reingold [26], which belongs to the class of force-directed

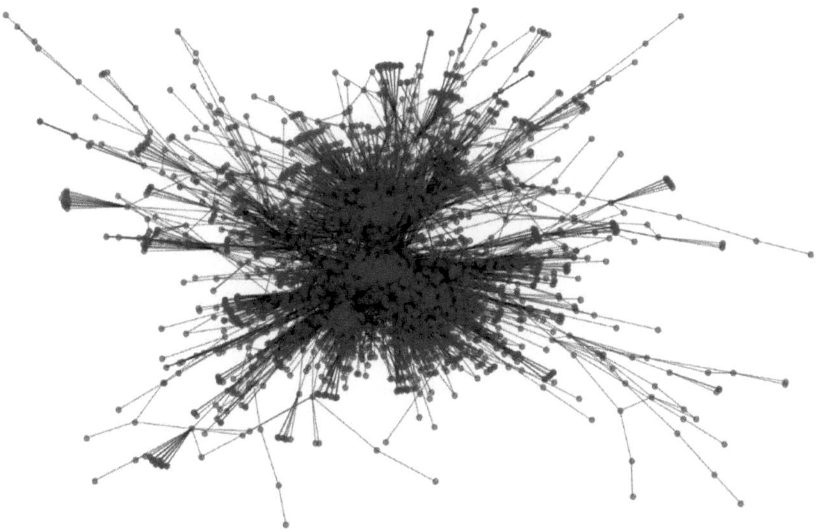

FIG. 9.2 Visualization of the global air transportation network using the force-directed algorithm Fruchtermann-Reingold, instead of geo-spatial information. *Notes:* Distances of links are minimized for the purpose of visualization. The figure exposes how several nodes aggregate into well-connected clusters. Moreover, it also exposes how certain nodes act as gatekeeper for the accessibility of other nodes to the network.

algorithms. In summary, force-directed algorithms aim to draw all the edges more or less equal length and there are as few crossing edges as possible. This visualization is interesting because it reveals a common pattern in air transportation, which is important for resilience. We can observe several fan-like substructures, where one airport controls access to the network for several other airports. All the airports behind the fan's main airport have to rely on the main airport; if the main airport fails, these other airports are immediately disconnected from the whole network. Moreover, we can see that several nodes in the network apparently fall into larger groups with high connectivity. We will further analyze this property in the following sections.

9.4 AIR TRANSPORTATION NETWORK ANALYSIS

Based on the complex network representation, we perform several analyses which focus on robustness estimation of the global air transportation network, including centrality measures (Section 9.4.1), efficiency of disrupted networks (Section 9.4.2), air/side accessibility of nodes (Section 9.4.3), community structures induced by highly-connected subgraphs (Section 9.4.4), airline network layers (Section 9.4.5), and Multiple Airport Regions (Section 9.4.6). With these (mostly) complex network tools, we have a means to analyze the resilience of the air transportation network from complementary perspectives.

9.4.1 Centralities

Recently, complex network researchers have used standard metrics to analyze networks, boosted by their success for the analysis of social networks. These network metrics compute a score for each node, which represents the relative importance of a node in the network. Since the publication of the seminal paper by Freeman [27], many network metrics have been proposed, capturing different aspects of networks. Each of these metrics can be exploited as a strategy for attacking a network, by removing most important nodes before less important ones. We summarize several commonly used network metrics as follows.

(1) Degree: $D_i = \sum_j a_{ij}$, where a_{ij} is the connection between node i and node j : $a_{ij} = 1$ when there is a connection existing; $a_{ij} = 0$ otherwise. For all pairs of nodes in the network, a_{ij} composes the adjacency matrix. Degree refers to the number of connections with other nodes.

(2) Betweenness centrality: $B_i = \sum_{s \neq t} \frac{\sigma_{st}(i)}{\sigma_{st}}$, where σ_{st} is the number of shortest paths going from node s to node t; $\sigma_{st}(i)$ is the number of shortest paths going from node s to node t and passing through node i . This metric indicates the number of shortest paths going through a node.

(3) Closeness centrality: $C_i = \frac{\sum_{j \in N, j \neq i} \sigma_{ij}}{(n-1)}$, where N is the set of all nodes in the network, n is the number of nodes, $\sigma_{ij}$is the shortest path between node i and node *j*. This metric is the average distance from a given starting node to all other nodes in the network.

(4) Eigenvector centrality: It is the eigenvector for the largest eigenvalue of the adjacency matrix. Nodes with high eigenvector centrality also connect to other nodes which have high eigenvector centrality.

(5) HITS (Hyperlink-Induced Topic Search) hub: Estimates how valuable information stored in one node is, based on links.

(6) Page rank: A ranking of the nodes based on the structure of the incoming links.

In Fig. 9.3, we show the top-X nodes according the network metric degree. The absolute degree of nodes is a measure for direct centrality, i.e., how many neighbors can be reached within one hop. While it is interesting to look at individual node degrees, another important facet is the distribution of degrees in the network. In many real-world networks, the degree distribution is highly skewed, usually in some way that low-degree nodes occur much more frequently than high-degree nodes.

In Fig. 9.4, we show the cumulative degree distribution of global air transportation networks. We can see that the frequency of nodes decreases fast with an increasing node degree. From an air transportation perspective, this means that only a very few airports in the network are highly connected and build the so-called core of the network. The other nodes, having a significantly smaller degree, rely on the core for transportation with few hops.

All the above listed metrics rank the importance of nodes regarding different facets of the network. Accordingly, different metrics identify different nodes as important ones. Next, we look at the airports that have the highest betweenness in the network. The betweenness centrality measures how many times a node is part of a shortest path between any two nodes in the network. In other words, how many times would a node be used for transportation with as few hops as possible. In Fig. 9.5, we report the top-X nodes according the network metric

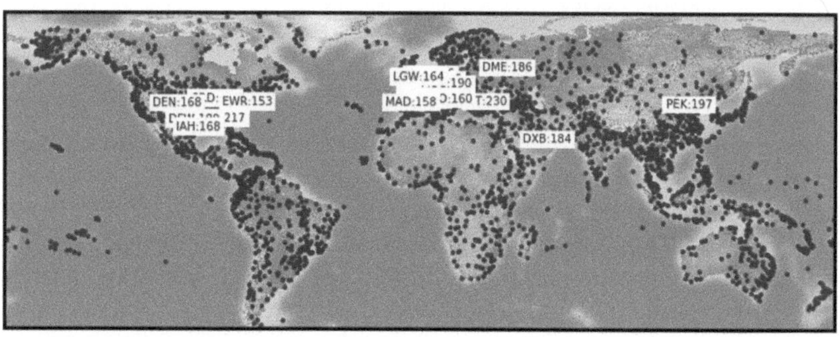

FIG. 9.3 Airports with Top-Degree values in global air transportation network. *Notes*: All airports are located in the northern hemisphere, with a strong focus on Western Europe and North America.

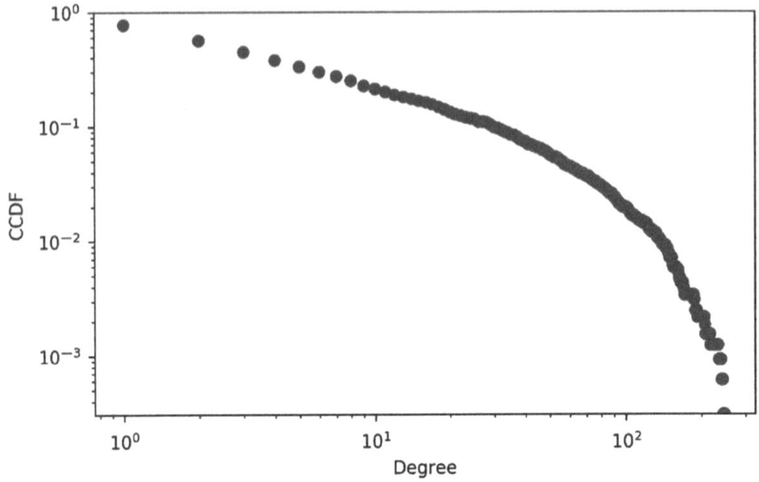

FIG. 9.4 Degree distribution for the global air transportation network. *Notes*: While nodes with low degree occur frequently in the network, the frequency of nodes with higher degree reduces fast. Only very few nodes have exceptionally high degrees. This structure gives the air transportation network its hub-and-spoke property.

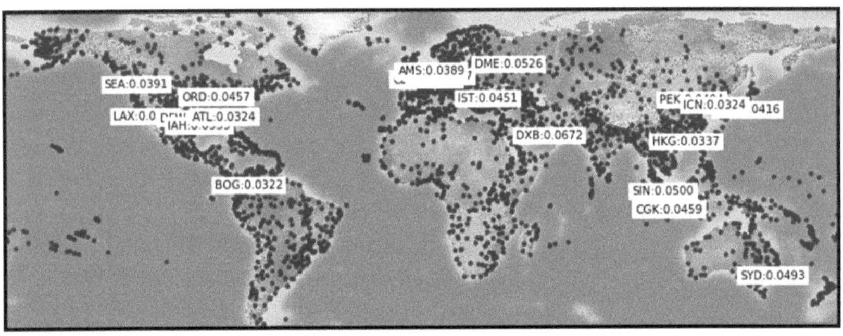

FIG. 9.5 Airports with Top-Betweenness values in global air transportation network. *Notes:* Most airports are located in the northern hemisphere. Compared to high-degree nodes, we also find important nodes in South Asia and Oceania.

betweenness. We are already familiar with some of the airports from the high-degree ranking, for instance, CDG (Charles de Gaulle Airport), IST (Istanbul Atatuerk Airport), DXB (Dubai International Airport), or PEK (Beijing Capital International Airport). Others appear to have a high betweenness centrality, without being highly connected, e.g., ANC (Ted Stevens Anchorage International Airport) and PPT (Fa'a'Ä International Airport). Such airports often control the access to other airports and are therefore critical for the resilience of their neighbors. In the case of ANC, many passengers traveling towards Alaska first go to ANC and afterwards switch to a local airline for their further travel.

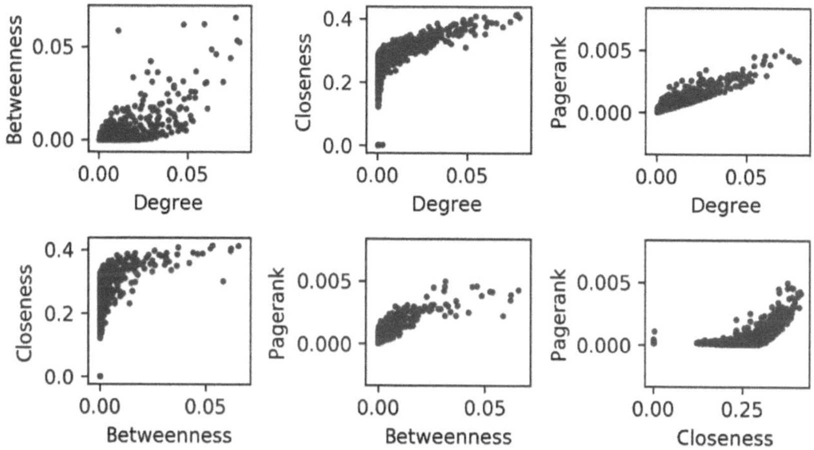

FIG. 9.6 Pairwise correlation of four centralities: degree, betweenness, closeness, and pagerank. *Notes:* We observe a weak correlation between most pairs only. Particularly, there is no strong correlation between degree and betweenness, which implies that high connectivity does not necessarily imply high throughput.

Next, we further investigate the correlation between selected metrics for global air transportation. We selected the following four metrics: Degree, betweenness, closeness, and pagerank. None of the metric pairs exhibits a strong correlation, yet, most of them are weakly correlated in a sense that nodes ranked high (low) against one metric are unlikely to be ranked low (high) in any other. These examples support the idea that the importance of a node cannot be represented by an individual metric only. Similarly, when analyzing the robustness of a network, different metrics often induce significantly different attacking strategies, with different observable effects on network robustness (Fig. 9.6). We evaluate these effects next.

9.4.2 Robustness Curves

As discussed in the previous section, network metrics induce different attacking strategies on a network. Therefore, studies often compare the results of different metrics when assessing the resilience of a network against targeted attacks. For estimating the effect of an attack to the network, one usually measures the reduction in network performance, expressed as the relative size of the largest component compared to the network without disruptions. Specifically, given a set of nodes, we divide the size of the giant component after removing the set of nodes by the size of the giant component before removing the nodes. The smaller the fraction, the more effective the attack on the network.

First, we investigate the robustness of global air transportation against random disruptions, where nodes are disrupted in no predefined order. The results are shown in Fig. 9.7. Global air transportation is robust against such kind of

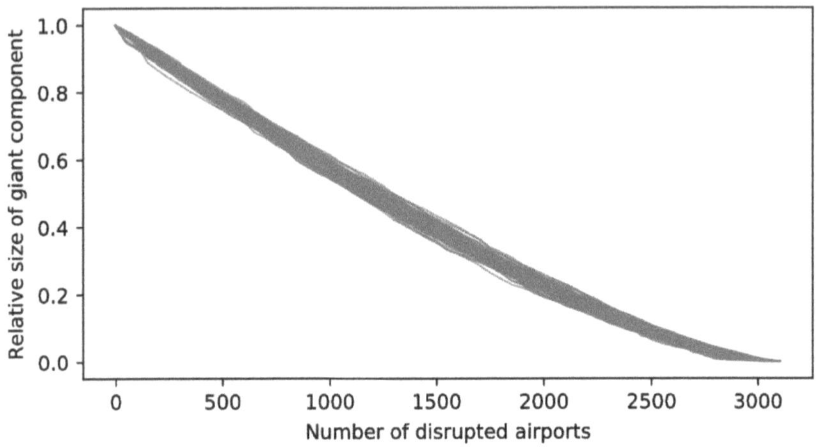

FIG. 9.7 Visualizing the relative size of the giant component under node removal according to 100 random attacks. *Notes*: Global air transportation is resilient against random attacks, as can be seen by the close-to-diagonal curves of random attacks.

random failures. Second, we assess the robustness of global air transportation against targeted attacks. We report the results of our experiments in Fig. 9.8. The global air transportation network is very vulnerable under all selected network metrics. If a few dozen nodes fail, the network's function is reduced to less than 90%. These central nodes are often responsible for large disruptions. Moreover, less than 20% of the nodes in the network are responsible for the complete functionality, indicated by reaching a relative size of the giant component close

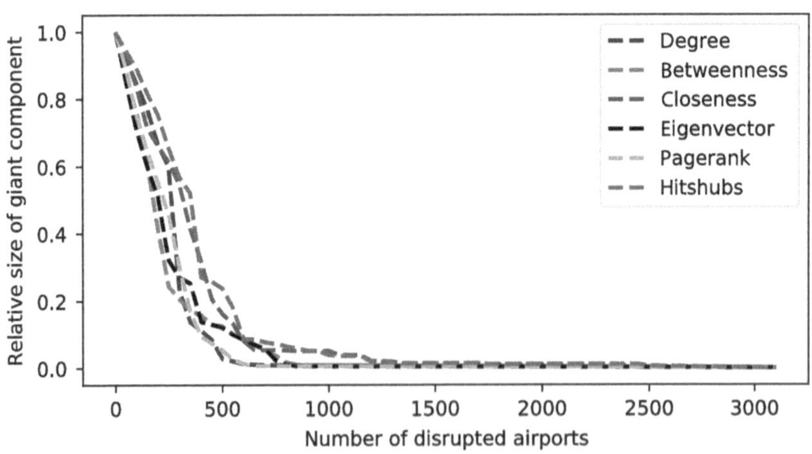

FIG. 9.8 Comparison of robustness curves, visualizing the relative size of the giant component under node removal according to different network metrics. *Notes:* Betweenness and eigenvector are the most effective attacking strategies for global air transportation.

to 0 for 500 airports. The betweenness-based attacking strategy is the most effective one for a large part of the attacking process; only when the network is almost fully destroyed is the degree-based strategy more effective.

9.4.3 Air-Side Accessibility of Nodes

Besides the topology-induced destruction of a network, we can also investigate the reachability of airports from other airports. The idea is to start with a particular node and plot the reachable nodes for a variable number of hops around that node, while the distance increases with the number of hops. Fig. 9.9 visualizes the hop-distance with six starting nodes. The nodes were selected according to their degrees. The selected airports are as follows:

- AMS: Amsterdam Airport Schiphol in Amsterdam, Netherlands
- TGD: Podgorica Airport in Podgorica, Montenegro
- SCN: Saarbruecken Airport in Saarbruecken, Germany
- LYC: Lycksele Airport in Lycksele, Sweden
- MFU: Mfuwe Airport in Mfuwe, Zambia
- OGD: Ogden-Hinckley Airport, Utah, United States

AMS, for instance, has the highest degree, while OGD has one of the lowest degrees in the network. We can see how most of the other airports can be reached from AMS with 1–2 hops; only a few of the 3000 airports need three hops or more. For these less connected airports, e.g., MFU and OGD, we can see that the majority of airports can only be reached with 3–5 hops.

9.4.4 Communities

In the previous sections, we have mostly reported the roles of individual airports in the network. One common analysis task for complex networks is to find out whether some parts of the network are locally better connected than others. In other words, can we identify groups of nodes whose internal connections are significantly different from their external ones? This concept is referred to as community structure in complex network research. A community is usually defined as a partition of the nodes, such that each node is put into only one community.

We show the communities of the global air transportation network in Fig. 9.10. The color of nodes indicates their community, i.e., nodes sharing the same color belong to the same community. Spatially-close airports often belong to the same community, with a few exceptions. The three largest communities are those for North America (763 airports), South-East Asia (583 airports), and Western Europe (505 airports). This shows that air transportation, despite being a long-distance travel mode, still is a strongly locality-based network, where spatially-close nodes are more likely to be connected.

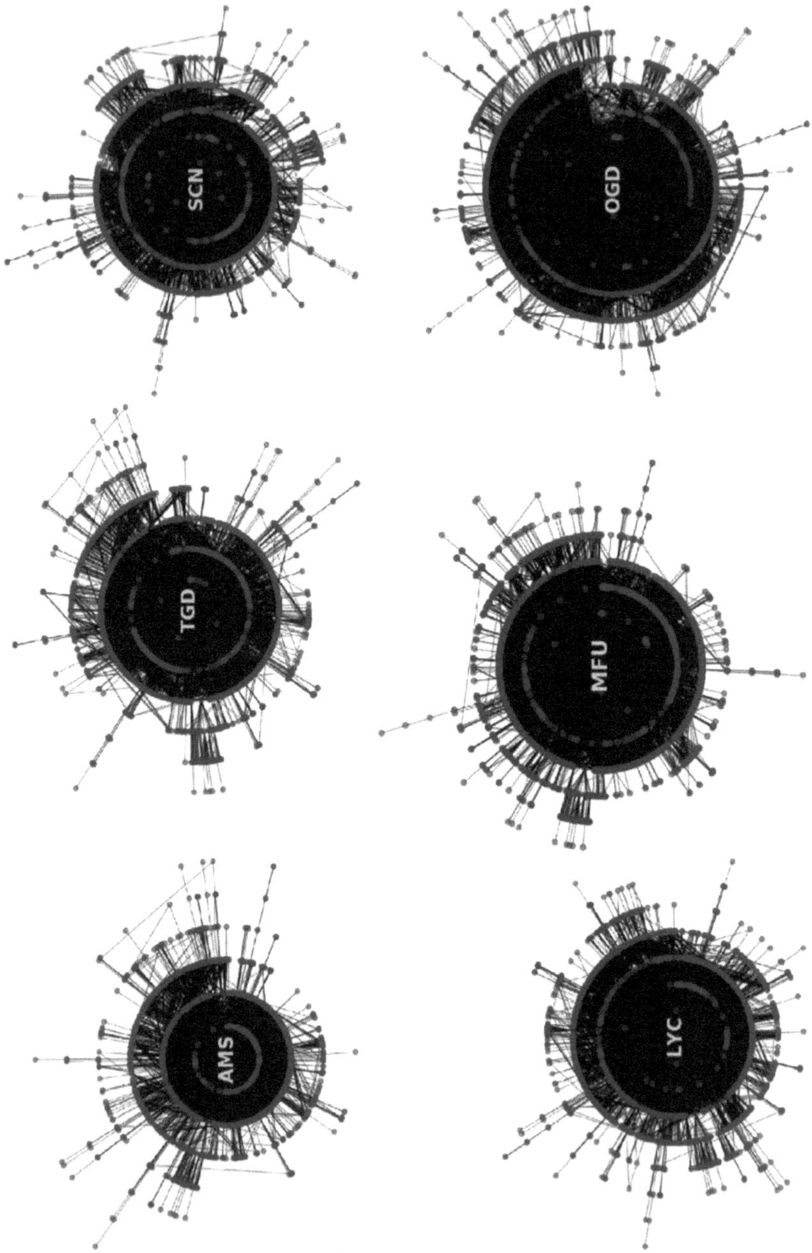

FIG. 9.9 Air-side accessibility of six airports in the global air transportation network. *Notes:* The source airports are labeled in the center with their IATA codes. The *concentric circles* report the reachability of airports with an increasing number of hops. Highly connected nodes, e.g., AMS (Amsterdam Airport Schiphol), are more accessible and closer to other airports than low-degree nodes, e.g., OGD (Ogden-Hinckley Airport, Utah, USA).

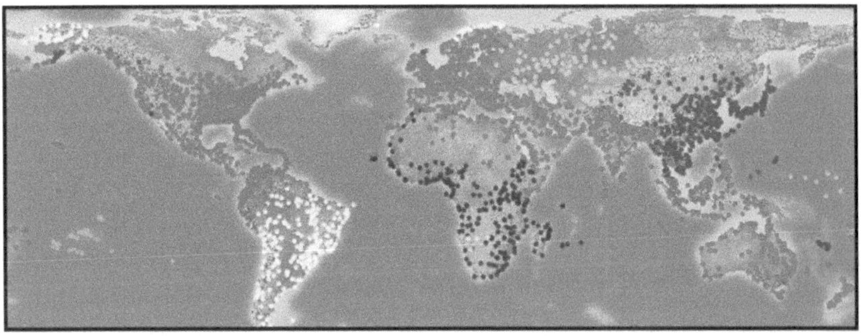

FIG. 9.10 Communities in the global air transportation network. *Notes*: Each color represents a different community. In total, we have 31 communities, where 4 communities cover approximately 60% of all airports. A clear spatially-induced distribution of communities can be observed.

9.4.5 Airline Networks

The global air transportation network is, in fact, composed of many layers. Each layer corresponds to a single airline. One interesting observation is that the airline networks—together spanning the global air transportation network—are surprisingly different in size and structure. We visualize two example networks, for Turkish Airlines (Fig. 9.11) and Ryanair (Fig. 9.12), respectively. The network of Turkish Airlines is clearly built around a few hubs in Central Turkey, supplying flights mainly from IST (Istanbul Atatuerk Airport) to other global destinations. The other airports in the Turkish airlines network have a significantly smaller degree than IST. This makes the network, which is referred to as hub-and-spoke structure, very vulnerable against disruptions at IST. Essentially, if IST is disrupted, the whole network will be affected. The second airline network, for Ryanair, is structured completely differently. The network largely follows a point-to-point pattern, where many nodes are connected by direct flights. This makes the network much more robust than the hub-and-spoke structure. Furthermore, we visualize the degree distributions of the

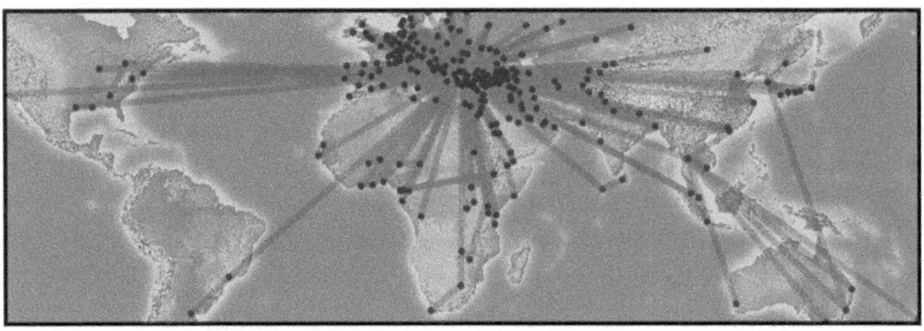

FIG. 9.11 Airline network of Turkish Airlines. *Notes*: The network covers a large number of international airports, almost all of them are operated from a single hub: IST (Istanbul Atatuerk Airport). A failure at IST is very likely to disrupt the whole network of Turkish Airlines.

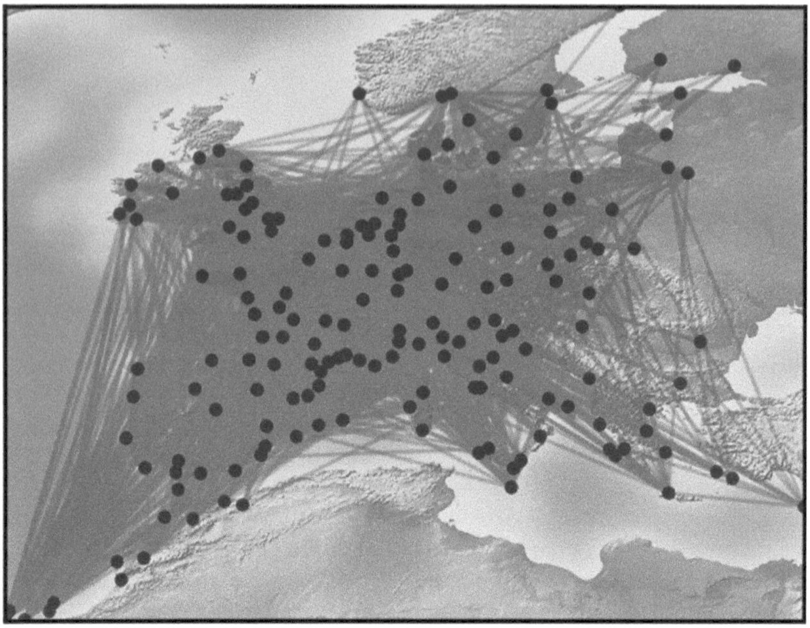

FIG. 9.12 Airline network of Ryanair. *Notes*: The network consists of many hub nodes and, accordingly, a failure at a single hub can often be compensated for by other airports.

two selected networks in Fig. 9.13. We can observe that the network of Turkish Airlines has very few high-degree nodes, while the network of Ryanair is less concentrated on a few hubs.

9.4.6 Multiple Airport Regions

Another facet of air transportation, relevant for estimating the robustness of the system, is the presence of many airports nearby each other. Analyzing the roles and functions of a single airport often provides a limited view on its real spatial importance, especially in metropolitan areas where several airports serve passengers with increasing long-distance mobility demands within a region. Therefore, Multiple Airport Regions (MARs) are an effective starting point for air transport research. The concept of MARs emerged in the 1990s, as a group of two or more major commercial airports in a metropolitan region. In Fig. 9.14, we show the major airports around the Greater London area. All these seven airports are within a driving distance of less than 2 h, a time acceptable for many passengers if they need to fly very long distance. Therefore, all airports compete for passengers in Greater London. On the other hand, in case of disruptions, airports can pick up the slack from a disrupted airport and help to increase the resilience of the air transportation system.

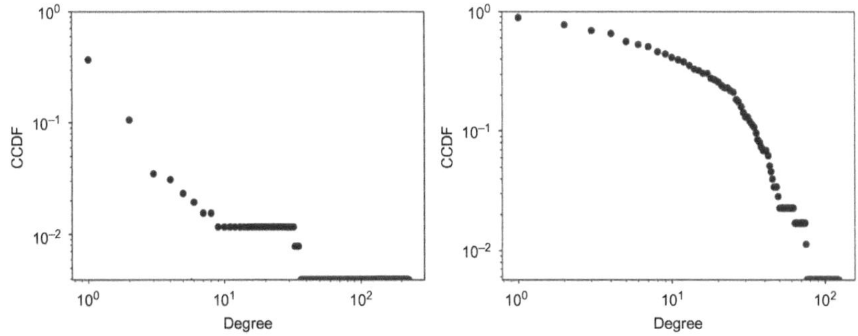

FIG. 9.13 Degree distribution for the airline networks of Turkish Airlines (left) and Ryanair (right). *Notes*: The left distribution has very few high-degree nodes, while the right degree distribution reveals less concentration on a few selected hubs.

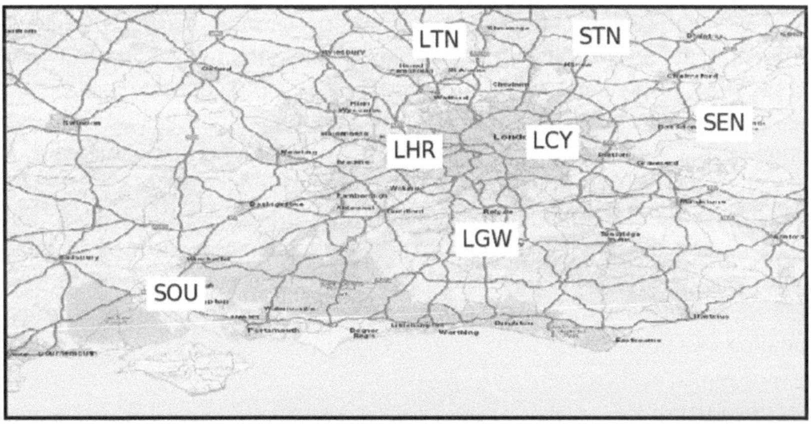

FIG. 9.14 An example of Multiple Airport Region (MAR) for the Greater London area. *Notes*: Seven airports serve the city, with different capacities, destinations, and accessibility.The methodology for computing MARs is usually based on spatial distances, often airports within 120–150 km. In Fig. 9.15, we visualize the global MARs which have at least five airports. Please note that, since openflights.org has no passenger data, the regions can contain airports with very little regular passenger traffic. We can see that the majority of MARs are found in Western Europe and North America. The air transportation subsystem in these areas is much more resilient than in other regions.

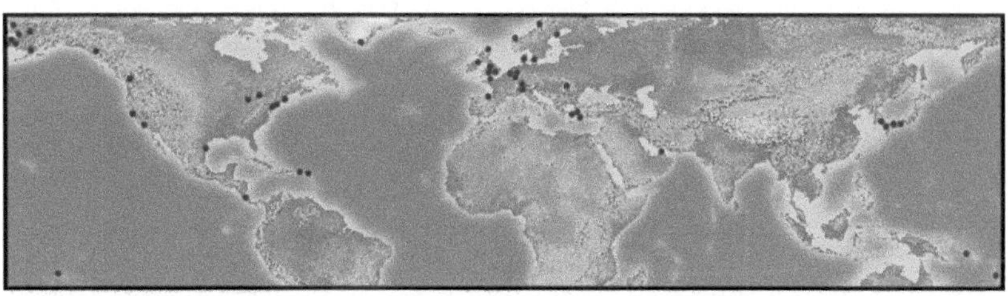

FIG. 9.15 Multiple Airport Regions (MARs) in the global airport network, with distance less than 120 km. *Notes*: Only MARs with at least five airports are shown. The majority of MARs are found in Western Europe and North America.

9.5 CONCLUSIONS

With the rapid development of economic globalization and increased demand for long-distance travel, air transportation is one of the critical infrastructure networks for a nation and has become the backbone of economies. Disruptions in air transportation, due to either extreme weather conditions, mechanical failures, or malicious attacks, can lead to huge economic losses. Therefore, in order to avoid such high socio-economic costs, it is critical to assess the effectiveness and resilience of air transportation systems against natural disasters or human-intended disruptions.

In this study, we provided a data-driven, open-source framework in order to assess the effectiveness and resilience of air transportation, with tools of Network Science and Data Science. First, we described how to model air transportation systems as complex networks, based on a reproducible workflow for data preparation. Second, several important properties of air transportation networks have been analyzed, including network centrality measures, the effectiveness of different network attacking strategies, community structures of the network, accessibility at the nodal level, different layers of airline networks, as well as the emerging concepts of Multiple Airport Regions. All of the analysis provides complementary perspectives on the resilience of air transportation networks.

Our proposed data-driven, open-source framework was based on freely available datasets from Openflights and it was implemented in Python. The datasets and codes have been made freely available for academic use. Our work contributes towards setting up a gold standard reference and pushing forward the state-of-the-art in the network modeling and resilience analysis of air transportation systems.

REFERENCES

[1] S. Wilkinson, S. Dunn, S. Ma, The vulnerability of the european air traffic network to spatial hazards, Nat. Hazards 60 (3) (2012) 1027–1036.

[2] R. Guimera, S. Mossa, A. Turtschi, L.A.N. Amaral, The worldwide air transportation network: anomalous centrality, community structure, and cities' global roles, PNAS 102 (22) (2005) 7794–7799.

[3] M. Zanin, F. Lillo, Modelling the air transport with complex networks: a short review, Eur. Phys. J. Spec. Top. 215 (2013) 5–21.

[4] A. Cook, H.A. Blom, F. Lillo, R.N. Mantegna, S. Micciche, D. Rivas, R. Vázquez, M. Zanin, Applying complexity science to air traffic management, J. Air Transp. Manag. 42 (2015) 149–158.

[5] X. Sun, S. Wandelt, F. Linke, Temporal evolution analysis of the european air transportation system: air navigation route network and airport network, Transportmetrica B: Transp. Dyn. 3 (2) (2015) 153–168.

[6] X. Sun, S. Wandelt, F. Linke, On the topology of air navigation route systems, in: Proceedings of the Institution of Civil Engineers-Transport, vol. 170, 2017, pp. 46–59.

[7] X. Sun, S. Wandelt, Network similarity analysis of air navigation route systems, Transport Res E-Log 70 (0) (2014) 416–434.

[8] S. Wandelt, X. Sun, J. Zhang, Evolution of domestic airport networks: a review and comparative analysis, Transportmetrica B: Transp. Dyn. (2017) 1–17. https://www.tandfonline.com/doi/full/10.1080/21680566.2017.1301274.

[9] O. Woolley-Meza, D. Grady, C. Thiemann, J.P. Bagrow, D. Brockmann, Eyjafjallajökull and 9/11: the impact of large-scale disasters on worldwide mobility, PLoS ONE 8 (8) (2013). 08.

[10] A. Cardillo, et al., Modeling the multi-layer nature of the european air transport network, Eur. Phys. J. Spec. Top. 215 (1) (2013) 23–33.

[11] V. Louzada, F. Daolio, H. Herrmann, M. Tomassini, Smart rewiring for network robustness, J. Complex Networks 1 (2013) 1–11.

[12] A. Marzuoli, et al., in: Resilience of the national airspace system structure: a data-driven network approach, Digital Avionics Systems Conference (DASC), 2014 IEEE/AIAA 33rd, 2014.

[13] S. Wandelt, X. Sun, X. Cao, Computationally efficient attack design for robustness analysis of air transportation networks, Transportmetrica A: Transp. Sci. 11 (10) (2015) 939–966.

[14] X. Sun, V. Gollnick, S. Wandelt, Robustness analysis metrics for worldwide airport network: a comprehensive study, Chin. J. Aeronaut. 30 (2) (2017) 500–512.

[15] Z. Neal, Refining the air traffic approach to city networks, Urban Stud. 47 (10) (2010) 2195–2215.

[16] S. Wandelt, X. Sun, Evolution of the international air transportation country network from 2002 to 2013, Transport Res E-Log 82 (2015) 55–78.

[17] X. Sun, S. Wandelt, M. Zanin, Worldwide air transportation networks: a matter of scale and fractality? Transportmetrica A: Transp. Sci. 13 (7) (2017) 1–24.

[18] R. de Neufville, Multi-airport systems in metropolitan regions, Tech. rep, Massachusetts Inst. Tech., 1986.

[19] A.B. Monteiro, M. Hansen, Improvements to airport ground access and behavior of multiple airport system: Bart extension to San Francisco International Airport, Transp. Res. Rec.: J.Transp. Res. Board 1562 (1996) 38–47.

[20] P. Bonnefoy, R. de Neufville, R. Hansman, Evolution and development of multiairport systems: worldwide perspective, J. Transp. Eng. 136 (11) (2010) 1021–1029.

[21] K. O'Connor, K. Fuellhart, Airports and regional air transport markets: a new perspective, J. Transp. Geogr. 53 (2016) 78–82.

[22] X. Sun, S. Wandelt, M. Hansen, A. Li, Multiple airport regions based on inter-airport temporal distances, Transport Res E-Log 101 (2017) 84–98.

[23] P. Fleurquin, J. Ramasco, V. Eguiluz, Systemic delay propagation in the US airport network, Sci. Rep. 3 (2013) 1–6.

[24] P. Baumgarten, R. Malina, A. Lange, The impact of hubbing concentration on flight delays within airline networks: an empirical analysis of the US domestic market, Transport Res E-Log 66 (0) (2014) 103–114.

[25] S. Belkoura, J.M. Peña, M. Zanin, Generation and recovery of airborne delays in air transport, Transp. Res. C Emerg. Technol. 69 (2016) 436–450.

[26] T.M. Fruchterman, E.M. Reingold, Graph drawing by force-directed placement, Softw.: Pract. Exp. 21 (11) (1991) 1129–1164.

[27] L.C. Freeman, Centrality in social networks: conceptual clarification, Soc. Networks 1 (1978) 215–239.

Chapter 10

Health Assessment of Electric Multiple Units

Tianyun Shi*, Haiyan Shen*, Li Li*, Peng Sun* and Ge Guo*,†
*Institute of Computing Technology, China Academy of Railway Sciences, Beijing, People's Republic of China, †Department of Civil and Environmental Engineering, University of Washington, Seattle, WA, United States

Chapter Outline

10.1 INTRODUCTION

Prognostics and Health Management (PHM) technology is a combination of advanced detection technology, diagnostic technology, and equipment maintenance management theory. Faults can be identified and dealt with based on massive data. At the same time, maintenance planning, guarantee decision-making, cost reduction, safety guarantee can be achieved. Finally, condition-based maintenance can be realized. Systematic PHM research and application of Electric Multiple Units (EMU) have not yet been carried out.

Chinese high-speed rail has achieved great development in recent years; however, there are still problems, such as vehicle equipment failure, design and manufacturing quality defects, and quality of maintenance defects, etc. The above-mentioned problems have unavoidable impacts on operation and maintenance of EMUs. At the same time, planned maintenance and corrective maintenance may result in excessive or ineffective maintenance, resulting in high maintenance costs, secondary disasters, and equipment failures. It is of great practical significance to establish a PHM system that is suitable for the conditions of high-speed railways in China.

Data-Driven Solutions to Transportation Problems. https://doi.org/10.1016/B978-0-12-817026-7.00011-4

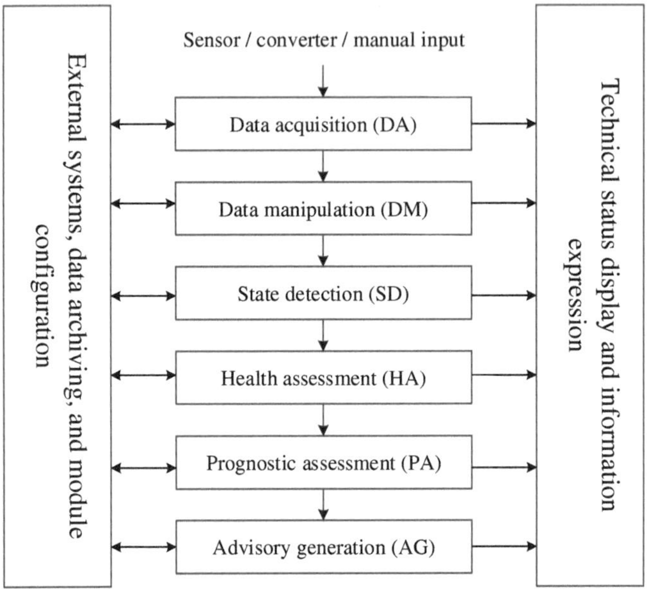

FIG. 10.1 ISO-13374 data processing and information flows [1].

In ISO 13374 in the CM&D standard, a PHM system consists of six functional modules is shown in the Fig. 10.1. The first three modules mainly obtain equipment status data through data collection, data processing. The latter three modules provide health assessments, warning signs, and suggestions (based on equipment status data, various diagnostic and prediction algorithms), evaluate equipment current health status, predict failures, and provide decision-making suggestions for troubleshooting. Health assessment is an important part of PHM.

In this article, the EMU's health assessment is implemented from the data point of view, using data fusion technologies [2–4]. Firstly, the structure and monitoring data of the EMUs are analyzed, and then the data fusion in the feature layer and the decision layer is carried out. Finally, the health index of the EMUs is obtained. It is proved that the health index can reflect the health status of EMUs; and can be used to support decision-making of EMUs' running and maintenance, and it has practical significance for the operation safety of EMUs.

At present, the assessment of health status is mainly concentrated in the fields of aerospace, shipbuilding, and electric power at home and abroad, based on methods including failure mode, effects, and criticality analysis (FMECA), failure mechanism analysis, and failure model analysis [5–8]. In the literature [5], a method of assessing the health status of unmanned aerial vehicle (UAV) systems based on the improved FMECA method is proposed. The health status of the UAV system is evaluated when failures occur and after the failures are handled. The assessment method is operational and easy to implement.

However, the algorithm is too complex, and efficiency is low, and it is evaluated after failures occur, which greatly reduces the significance of the evaluation. In the literature [6], health status parameters are classified based on the grading evaluation method, and the residual life is evaluated based on health index theory. Correction factors of health status parameters are obtained based on design life, not on operation status, which would affect assessment accuracy, and this method is limited to oil insulation equipment.

In the literature [7], the index of uncertainty and importance degree are introduced according to the health status assessment problem of lower accuracy and objectivity, and a method of health status assessment based on uncertainty and importance degree using an improved Dezert-Smarandache Theory (DSmT) is proposed. The evaluation results are verified and have a high reference value. In the literature [8], a quantitative assessment method of rolling bearing health status is proposed combining Chaos Fruit Fly Optimization Algorithm (CFOA) with Multikernel Hypersphere Support Vector Machine (MKHSVM) to achieve the health status quantitative assessment of a rolling bearing. The above algorithms are too complex, consume too much computing time and resources, and there are too many subjective factors.

Based on the PHM open system, a data fusion method is proposed to implement EMU health status assessment. There are many common data fusion methods including:

- HIS transform, PCA transform, wavelet transform, weighted average, and other classic methods, which are suitable for data-level data fusion mainly used in multisource image composition, image analysis and understanding;
- cluster analysis, Bayes estimation method, information entropy method, weighted average method, DS evidence reasoning method, voting method, and neural network method, which are suitable for feature layer data fusion, mainly used for multisensor target tracking, parameter correlation analysis, and state vector estimation, etc.;
- Estimation method, expert system, neural network method, fuzzy set theory, reliability theory, and logic template method, which are suitable for decision-level data fusion [9], and can provide support for control command and decision.

10.2 DATA SOURCE AND STRUCTURE

In order to support modeling and calculation in PHM of EMU, and to implement prognostics and health assessment applications, various data from different sources are needed for storage, screening, conversion, and cleaning, including monitoring data, inspection data, manufacturing data, environmental data, and maintenance data.

The on-board monitoring system of EMUs is relatively mature. There are hundreds of sensors on an EMU, collecting various data continuously while the EMU is running, and a huge amount of data has been accumulated that can support data-driven health assessment. The on-board monitoring system is the main testing form of Chinese EMU. The high-speed EMU train network control system is used as the nerve center system of EMU to implement train control, monitoring, diagnosis, and protection tasks. It is a core system and key technology of EMU. The EMU network control system mainly includes a central control unit, human-machine interface display, input and output workstations, input and output modules, temperature acquisition module, remote wireless transmission device, gateway, repeater, and so on. Fault diagnosis and prediction are one of the main functions of the network control system. Monitoring strategies are different for different EMU models; there are hundreds of sensors in the entire vehicle and more than 100 in the carriage. Sensor distribution of one EMU model is shown in Fig. 10.2.

Taking the example of the 8-formation EMU, the information acquisition of the current network control system divides the EMU's sensors into the categories of running gears, power systems, carriers, and basic information. The specific descriptions of various kinds of sensor data for EMU are as follows:

- High-voltage monitoring: the control and status of high-voltage system equipment such as pantographs, main circuit breakers, high-voltage isolation switches, network pressure sensors, network flow sensors, and main transformers.
- Traction converter: working status, fault information for communication and power supply.

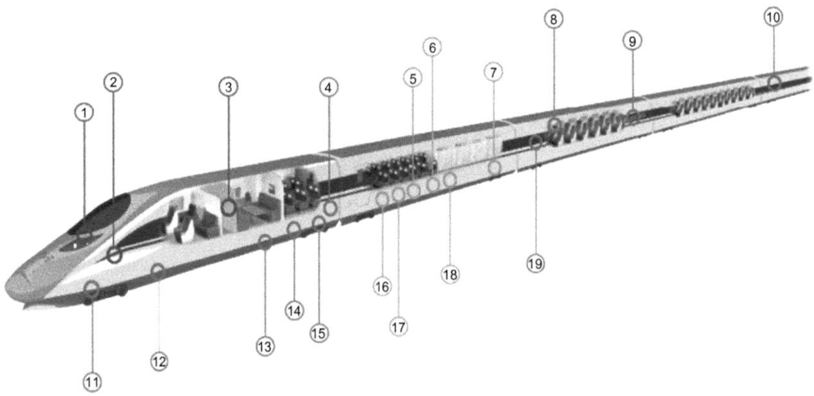

FIG. 10.2 Sensor distribution. 1: car information controlling device display screen, 2: cab temperature sensor, 3: wireless data transmission device, 4: external temperature sensor, 5: traction transformer oil flow device, 6: traction converter current/voltage sensor, 7: motor temperature sensor, 8: passenger car temperature sensor, 9: smoke and fire alarm probe, 10: net pressure transformer, 11: ATP speed sensor, 12: brake speed sensor, 13: semi active control acceleration sensor, 14: axis temperature sensor, 15: acceleration sensor for bogie instability detection, 16: overvoltage/lightning protection, 17: traction transformer primary current sensor, 18: brake control device pressure sensor, 19: car door sensor.

- Traction motors: shaft temperature, transmission end bearing temperature, nondrive end bearings temperature, and warning and alarm information.
- Auxiliary system: comparing the speed and temperature values with the set value to control cooling fan operation.
- Transformer cooling fan: vehicle speed and traction transformer temperature.
- Converter cooling fan: vehicle speed and traction converter temperature.
- Motor cooling fan: vehicle speed and traction motor temperature.
- Charger monitoring: diagnostic information of abnormal operation of charger internal devices, and battery power.
- Braking system monitoring: communication, command, power distribution results, and braking and safety loops.
- Axle temperature monitoring: TDD collects bogies and traction drive system bearings and motor temperature, with early warning, alarm, speed limit function, and fault diagnosis function.
- Door monitoring: feedback to open, close, isolate, fault, release, etc.
- Air-conditioning monitoring: indoor and outdoor temperature air-conditioning real-time work mode.
- Bogie instability monitoring: instability alarm information, and speed limit.
- Fire Alarm: Alarm and Fault Information, and Air Conditioning Shutdown.
- Emergency brake warning for passengers.

The data level fusion of EMU monitoring information is carried out mainly in the monitoring equipment. As the EMU's monitoring data can be obtained directly from the existing monitoring information systems, the data layer fusion study is not implemented in this article. When the same monitoring object has multiple sensors to monitor the same parameter, it is suitable to adopt the layer fusion feature. When the health status of the integrated system is evaluated by the fusion health status of multiple parts, decision layer fusion is needed. The data fusion strategy is shown in Fig. 10.3.

It's necessary to carry out data fusion for health assessment of EMU. According to the EMU's system structure proposed in the previous section, the health status assessment is divided into the component level, system level, car level, and train level. Integrating the monitoring data of different components allows comprehensive evaluation of the health status of the system, integrating the monitoring data or assessment results of different systems allows comprehensive evaluation of the health status of the car, and integrating assessment results of different cars allows comprehensive evaluation of the health status of the EMU. One object can have multiple types of monitoring, for example, EMU axles. Temperature sensors are used in the network control system to monitor axle temperature in real time, infrared detection equipment is used in ground monitoring systems to monitor axle temperature, acoustic sensor arrays are used in ground monitoring systems to monitor their noise; so a variety of monitoring is needed for data fusion to comprehensively evaluate the health

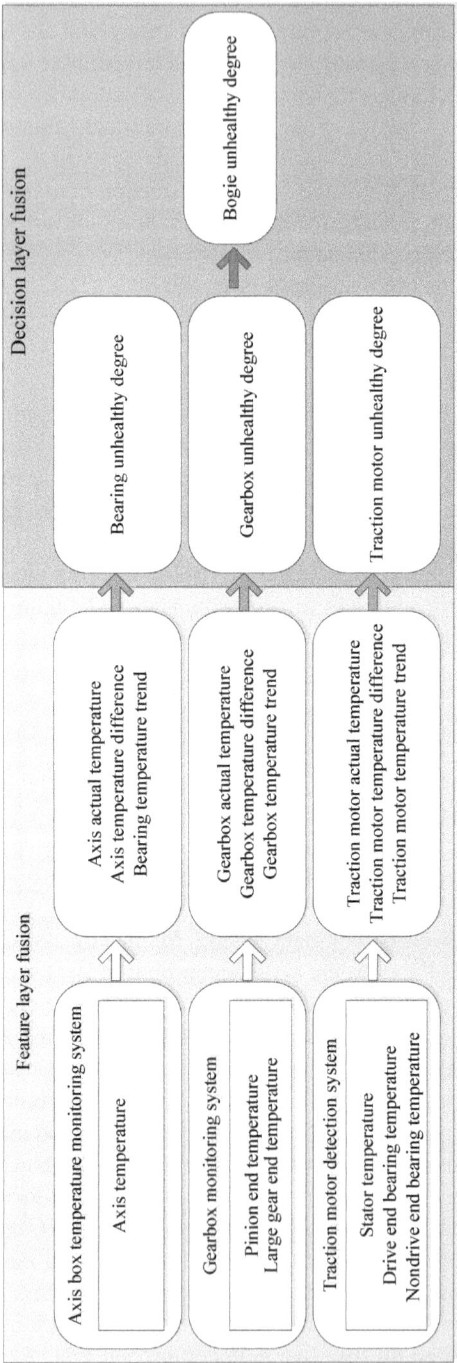

FIG. 10.3 Data sources and their fusion processing.

status of axle bearings and axle boxes. Moreover, closely related components have a significant impact on the health status of each other, for example, track defects would influence wheelsets and wheelset defects would influence axle and bearings, etc. As a result, it's necessary to integrate monitoring data of closely related components as an important factor in evaluating the health status of components.

10.3 HEALTH ASSESSMENT OF EMU

The objects of EMU direct monitoring are components; to assess the EMU's system-level components it is necessary to assess the health status of different parts of the system. EMU bogie systems consist of frame composition, wheel bearing boxes, primary suspension, secondary suspension, gearbox, and bogie accessory. Moving parts, including wheel bearing boxes and gearbox, have high failure occurrence rate and have a great impact on the health of the bogie. In addition, the abnormal vibration of the traction motor bearing installed in the bogie will cause bogie abnormalities even in the presence of vibration damping devices, and then affect bogie health status. In summary, taking into account the availability and reliability of monitoring projects and data, the bogie shaft temperature, gearbox temperature, and the traction motor temperature are taken as bogie health status assessment indicators.

10.3.1 Feature Layer Data Fusion

The gearbox sensor measures the temperature of the large gear and the small gear, and the traction motor sensor measures the temperature of the stator and the bearing at both ends. All of them measure the same parameters at different positions. Therefore, the feature layer data fusion selects the fusion of measured same values in different positions. The processing algorithm is the mean-based batch estimation theory [10–12]. The principle is that for a sensor of the same type, a set of measurement data is obtained first, and then the two sensors adjacent to each other in space are divided into two groups for calculation, based on the principle that they are not in one group. The algorithm is as follows.

Suppose there are k sensors, divided into two groups based on the principle that two adjacent sensors are not in one group. The first set of measurement data is $X_{11}, X_{12}, \ldots, X_{1m}$, the second set of measurement data is $X_{21}, X_{22}, \ldots, X_{2n}$, and "$m+n=k$." The arithmetic mean and standard deviation of the two groups are respectively:

$$\overline{X_1}=\frac{1}{m}\sum_{i=1}^{m}X_{1i}\sigma_1=\sqrt{\frac{1}{m-1}\sum_{i=1}^{m}\left(X_{1i}-\overline{X_1}\right)^2},\overline{X_2}=\frac{1}{n}\sum_{j=1}^{n}X_{2j},$$

$$\sigma_2=\sqrt{\frac{1}{n-1}\sum_{j=1}^{n}\left(X_{2j}-\overline{X_2}\right)^2}$$

σ^- and X^- indicate the standard deviation and fusion result of the last measurement. σ^+ and X^+ indicate the current measurement standard deviation and fusion results. The variance after processing is:

$$\sigma^+ = \left[(\sigma^-)^{-1} \cdot H^T C^{-1} H\right]^{-1} \tag{10.1}$$

In the formula, H is the coefficient matrix of the measurement equation:

$$H = \begin{pmatrix} 1 \\ 1 \end{pmatrix}$$

C is the covariance of the measurement noise:

$$C = E\left[V\ V^T\right] = \begin{bmatrix} \sigma_1^2 & 0 \\ 0 & \sigma_2^2 \end{bmatrix}$$

brought into Eq. (10.1):

$$\sigma^+ = \frac{\sigma_1^2 \sigma_2^2}{\sigma_1^2 + \sigma_2^2} \tag{10.2}$$

$$X^+ = \left[\sigma^+ \cdot \left(\sigma^{-1}\right)^{-1}\right] X^+ + \left[\sigma^+ \cdot H^T C^{-1}\right] \begin{pmatrix} \overline{X_1} \\ \overline{X_2} \end{pmatrix}$$

and then get the fusion result:

$$X^+ = \frac{\sigma_1^2}{\sigma_1^2 + \sigma_2^2} \overline{X_1} + \frac{\sigma_2^2}{\sigma_1^2 + \sigma_2^2} \overline{X_2} \tag{10.3}$$

This is the time when the next batch of data arrives and X^+ is σ^- and X^- for the next time data fusion. The initial values of σ^- and X^- are respectively ∞ and the average of the current data.

10.3.2 Decision-Making Level Data Fusion

When a monitoring object has multiple monitoring systems or an object is composed of multiple components, the comprehensively evaluated health status of the monitoring object and data fusion need to have good timeliness and fault tolerance to conduct decision-making level data fusion. Firstly, the weighted voting method [13] was adopted for decision-making level fusion. The monitoring of the bearing box, gearbox, and traction motor was regarded as the fusion of the three monitoring system data, and the final result of the EMU health assessment was obtained. The degree of system malady is used to reflect the degree of health, that is, the degree of system defect is described by the degradation of the system status or the fault condition—the higher the degree of maladjustment, the smaller the health index.

The weighted voting method can be described as having $N + 1$ thresholds, one for each monitoring system and for the required number of votes. If the monitoring system monitoring quantity is greater than the given threshold, the voting supports the monitoring system. When the number of votes for a decision is greater than a given threshold, this decision is supported. The data fusion value of the measurement amount of the ith monitoring subsystem is denoted by M_i, and T_i represents the threshold of the ith monitoring subsystem. Assuming that there are a total of N monitoring systems, the determination of monitoring results by the ith monitoring system can be expressed as:

$$R_i = \begin{cases} 1, M_i > T_i \\ 0, M_i \le T_i \end{cases} \tag{10.4}$$

N monitoring subsystem monitoring results:

$$\sum_{i=1}^{N} R_i > T \tag{10.5}$$

T is the total threshold of the decision. Considering that different monitoring systems have different effects on the discrimination result, due to the different performance of the monitoring system and the different environment of the received signal during the fusion process of different monitoring systems, the decision results should tend to contribute to a larger system. Different systems reflect different aspects of the objects to be evaluated, so that after they are united, more judgment information can be obtained and the judgment result is "more" credible. To deal with the above situation, a weighted vote is merged into:

$$\omega_1 R_1 + \omega_2 R_2 + \cdots + \omega_N R_N > T \tag{10.6}$$

ω_i is the weight value of the ith monitoring subsystem. The influence degree of the selected three evaluation indicators on the health status of the EMU is different. In the fusion process, the monitoring data of each component need to be weighted. Due to the characteristics of the use and maintenance of EMUs, the algorithm requires high efficiency and high stability. The widely used methods of adaptive learning in other fields are not suitable for obtaining individual weights. Several systems jointly make the final decision; the weight assigned to each system should reflect the importance of the system's role in the final decision. Obviously, the joint detection of multiple monitoring systems is more accurate than that of separate monitoring systems, and the misjudgment of missed sentences will be greatly reduced. Therefore, this multisystem cooperation can be regarded as an N-person cooperation strategy model.

N individuals engage in certain activities, for each type of cooperation among several of them (a single person is also regarded as a kind of cooperation), they will receive certain benefits. When the interests of people are nonadversarial, the increasing number of people in cooperation will not result in a reduction in the benefits. The cooperation of all N individuals will bring the greatest benefit. Thus the collection of N individuals and the benefits of cooperation of various forms constitute the N-person cooperative games. The Shapley value [14] is a scheme to distribute this maximum benefit. Its definition is as follows.

Let the set $I = \{1,2,\ldots,n\}$, if any subset s for I corresponds to a real-valued function $v(s)$ that satisfies $v(\phi) = 0, v(s_1 \cup s_2) \geq v(s_1) + v(s_2)$, $s_1 \cap s_2 = \phi$, say $[I, v]$ N cooperation strategy, v is the characteristic function of the strategy.

Use x_i to represent the amount of income that member i of group I should receive from the maximum benefit of cooperation $y(I)$. $x = (x_1, x_2, \ldots, x_n)$ is called the distribution of cooperation measures, and this distribution satisfies:

$$\sum_{i=1}^{N} x_i = v(I), x_i \geq v_i, \quad i = 1,2,\ldots,N.$$

The Shapley value is determined by the characteristic function v and is noted as$\phi(v) = (\varphi_1(v), \varphi_2(v), \ldots, \varphi_N(v))$, is a specific allocation, which is $\varphi_i(v) = x_i$, where

$$\varphi_i(v) = \sum_{i \in s} \omega(|s|)[v(s) - v(si)], \quad i = 1,2,\ldots N \tag{10.7}$$

$$\omega(|s|) = \frac{(N - |s|)!(|s| - 1)!}{N!}$$

where $|s|$ the number of elements in s is, $\omega(|s|)$ is weighting factor.

10.4 DATA APPLICATION AND ANALYSIS

10.4.1 Feature Layer Health Data Analysis

Long-term field application experience shows that the measured temperature can reflect the abnormal operation of the components, but the same side temperature difference (the difference between the left or right position of the same car in the forward direction) and the same location temperature difference (the same position of different trains in the same train). The difference can eliminate the effect of day and night temperature difference and sunlight, and reflect the abnormality of parts more accurately. The difference between the temperature and position is selected for evaluation. Therefore, based on the above-mentioned average value-based batch estimation theory, the temperature, temperature difference, and temperature change trend of the gearbox and the traction motor undergo data fusion (the axis temperature has only one measurement point and no data fusion is needed—Fig. 10.4 shows the measured values). Therefore, the measurement temperature and temperature difference of the two components are respectively merged, and the fusion results are shown in Figs. 10.5 and 10.6. The top of the three graphs is the temperature value, and the lower graph is the temperature difference.

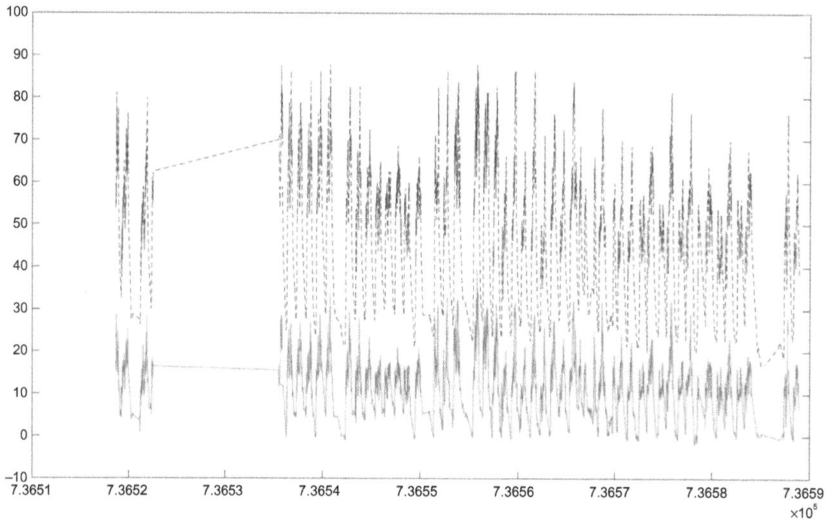

FIG. 10.4 Axis temperature and its difference.

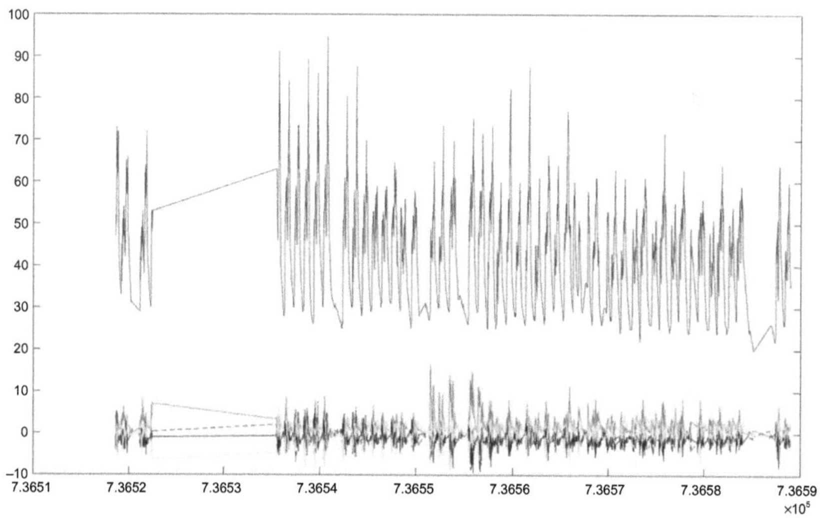

FIG. 10.5 Gearbox temperature and difference fusion result.

In addition, real-time tracking of temperature trends, and timely detection of continuous temperature rise phenomenon has an important role in predicting faults before the temperature reaches the upper limit temperature alarm trigger. In this paper, the slope of the envelope of the temperature versus position difference curve is used to represent the temperature trend.

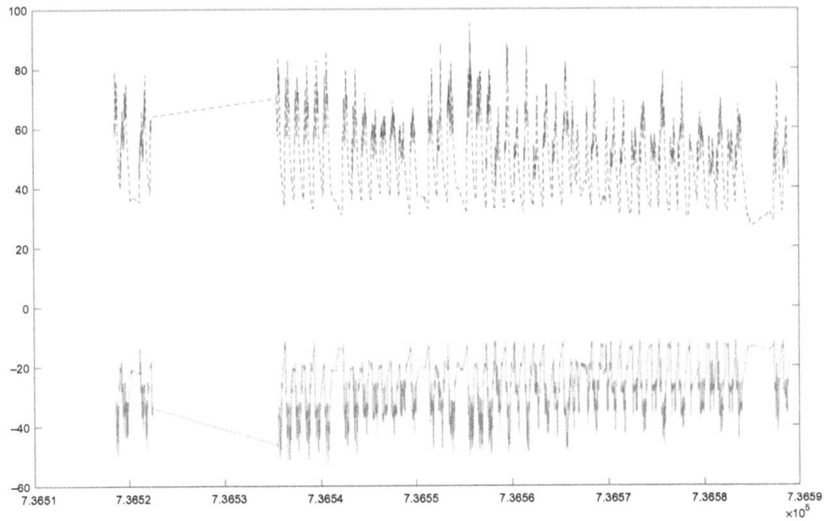

FIG. 10.6 Traction motor temperature and difference fusion results.

10.4.2 Decision-Making Level Health Data Analysis

The results of the fusion of the characteristic layers of the bearing housing, gearbox, and traction motor (including temperature, difference, and change trend) are regarded as three monitoring system monitoring data. The three monitoring system temperature monitoring systems, differential monitoring systems, and change trend monitoring systems are system 1, system 2, and system 3, respectively, which define the temperature, difference, and change trend-monitoring system's contribution to the health assessment of each component based on on-site expert experience. If a single monitoring system is used to assess the degree of component failure, the contribution of System 1 is 75%, the contribution of System 2 is 80%, and the contribution of System 3 is 70%. In the joint assessment, the degree of contribution of joint evaluation of systems 1 and 2 is 90%; the contribution of joint evaluation of systems 1 and 3 is 80%; and the contribution of joint assessment of systems 2 and 3 is 85%. The joint evaluation of the three systems contributes 95%. Mathematical modeling and data processing are the same as those below. Due to space limitations, they are not described here. The result of the fusion is the defective degree R of the components, as shown in Fig. 10.7, where the blue graph shows the defective degree of the bearing box, the red graph shows the defective degree of the gearbox, and the green graph shows the defective degree of the traction motor.

The monitoring system of the three monitoring systems of the EMUs, the bearing box monitoring system, the gearbox monitoring system, and the traction motor monitoring system are system 1, system 2, and system 3, respectively. The contribution of the axle, gearbox, and traction motor monitoring system

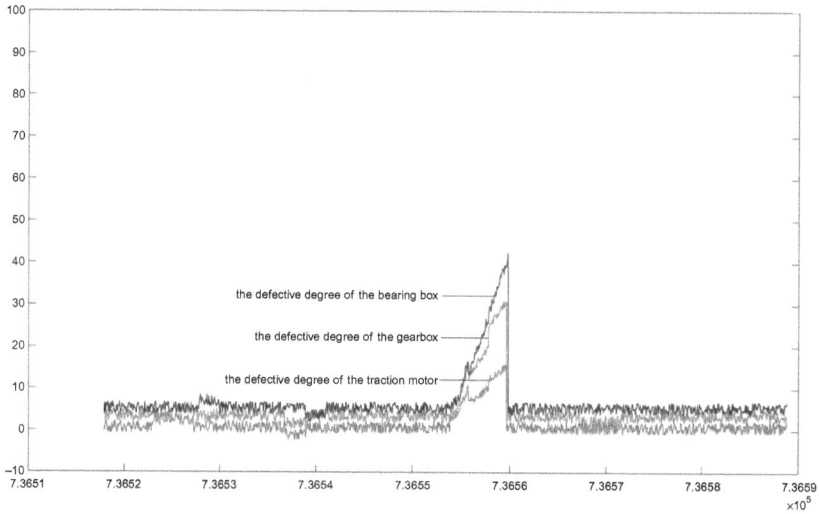

FIG. 10.7 Defective degree of bearing box, gearbox, and traction motor.

to the EMUs health assessment is defined according to the experience of on-site experts. If a single monitoring system is used to assess the health status of an EMU, the contribution of system 1 is 60%, the contribution of system 2 is 40%, and the contribution of system 3 is 20%. In the joint assessment, the degree of contribution of joint evaluation of systems 1 and 2 is 76%; the contribution of joint evaluation of systems 1 and 3 is 68%; the contribution of joint evaluation of systems 2 and 3 is 45%. The joint evaluation of the three systems contributes 81%.

Then the set $I=\{1,2,3\}$ and the evaluation contribution is defined as the characteristic function on I, i.e.:

$$v(\phi)=0, v(1)=60\%, v(2)=40\%, v(3)=20\%, v(1\cup 2)=76\%, v(1\cup 3) \\ =68\%, v(2\cup 3)=45\%, v(I)=81\%$$

then, the proportions of the calculation system 1, system 2, and system 3 in weighted voting are calculated using Tables 10.1–10.3.

Substituting the data of the last row in Tables 10.1–10.3 into Eq. (10.7):

$$\varphi_1(v)=46\%, \varphi_2(v)=24.5\%, \varphi_3(v)=10.5\%.$$

Thus, in formula (10.6):

$$\omega_1=\frac{46\%}{81\%}=0.57, \omega_2=\frac{24.5\%}{81\%}=0.3, \omega_3=\frac{10.5\%}{81\%}=0.13$$

substituted into Eq. (10.6) the defective degree f of the EMUs is:

$$f=0.57R_1+0.3R_2+0.13R_3 \tag{10.9}$$

TABLE 10.1 Contribution of System 1 in System Joint

s	1	$(1 \cup 2)$	$(1 \cup 3)$	I
$v(s)$	60%	76%	68%	81%
$v(s\backslash 1)$	0	40%	20%	45%
$v(s) - v(s\backslash 1)$	60%	36%	48%	36%
$\vert s\vert$	1	2	2	3
$\omega(\vert s\vert)$	1/3	1/6	1/6	1/3
$\omega(\vert s\vert)[v(s) - v(s\backslash 1)]$	20%	6%	8%	12%

TABLE 10.2 Contribution of System 2 in System Joint

s	2	$(1 \cup 2)$	$(2 \cup 3)$	I
$v(s)$	40%	76%	45%	81%
$v(s\backslash 1)$	0	60%	20%	68%
$v(s) - v(s\backslash 1)$	40%	16%	25%	13%
$\vert s\vert$	1	2	2	3
$\omega(\vert s\vert)$	1/3	1/6	1/6	1/3
$\omega(\vert s\vert)[v(s) - v(s\backslash 1)]$	13.3%	2.7%	4.2%	4.3%

TABLE 10.3 Contribution of System 3 in System Joint

s	3	$(1 \cup 3)$	$(2 \cup 3)$	I
$v(s)$	20%	68%	45%	81%
$v(s\backslash 1)$	0	60%	40%	76%
$v(s) - v(s\backslash 1)$	20%	8%	5%	5%
$\vert s\vert$	1	2	2	3
$\omega(\vert s\vert)$	1/3	1/6	1/6	1/3
$\omega(\vert s\vert)[v(s) - v(s\backslash 1)]$	6.7%	1.3%	0.8%	1.7%

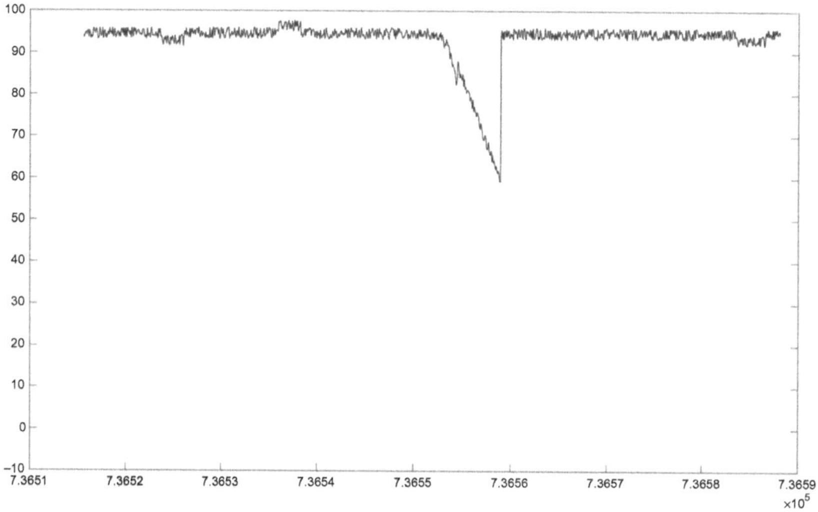

FIG. 10.8 EMU's health index.

Define the EMUs health index $h = 1 - f$. Obviously, when f is greater, the worsening degree of the EMUs is more serious, the health index is smaller, and the health status is worse. Divide the f-value into three intervals.

$$h \in \begin{cases} [0.85, 1], f \in [0, 0.15], \text{good} \\ [0.60, 0.85), f \in (0.15, 0.4], \text{normal} \\ [0, 0.6), f \in [0, 0.15], \text{severe} \end{cases} \tag{10.10}$$

When $h \in [0.85,1]$, indicating that the health status is good, no measures need to be taken; when $h \in [0.60,0.85]$, the health status is normal, the car group needs to be set as the key arc group, and the focus should be on the control. When $h \in [0,0.6]$, the status is severe, immediate measures need to be taken to manually confirm and, if necessary, take measures such as speed limits or even stopping.

The result of data fusion using the above method is the EMU's health index, as shown in Fig. 10.8.

It can be seen from Figs. 10.4–10.6 that the temperature of the bearing box, gearbox, and traction motor has exceeded the standard many times in mid-August. Corresponding to Fig. 10.7, it can be seen that the bad index of the bearing box, gearbox and traction motor rose significantly during this period, while the health index in Fig. 10.8 dropped sharply to around 0.6 during this period. In combination with the car groups fault data, it was found that the bearing had cracked after this period of time. The method was further verified by using the 3-month operating data and maintenance data of an EMU. During this period, the health index had a more obvious decline during the period of the fault that was found to be a safety hazard, and within a few days before the failure occurred, the decline in the health index was found, and the accuracy of the assessment reached 74%. This shows that the data-driven health assessment method proposed above can realize the assessment of the health status of EMU and can predict the health development trend to some extent.

10.5 CONCLUSION AND OUTLOOK

The EMU PHM is a hot spot for EMU to apply maintenance quality management and technical management work, which can effectively reduce the risk caused by faults in the EMU's operations and improve operational efficiency. Combining with the actual situation of the China Railway Corp. and Railway Bureaus, through the analysis of the needs of users, data, and functions, a complete set of EMU PHM system construction plans have been formed. This will improve the implementation of EMU train maintenance information systems, realize EMU condition-based maintenance, and improve the efficiency of EMUs' operations and maintenance.

However, in addition to the system construction itself, for the establishment of a complete PHM system, the following two parts should be completed simultaneously. First, continuous optimization of EMU configuration data. Configuration management can control changes in various configuration components and ensure the integrity and traceability of the component throughout the life cycle of the system, which is crucial for studying the reliability of components and even systems and vehicle groups. Second, the establishment of the EMU's health system is a complex and large project, the determination of each link requires a lot of theoretical and practical arguments. The EMU's health status should have a complete set of rigorous evaluation criteria.

Based on the analysis of the EMU structure and monitoring status above, a health assessment method based on data fusion is proposed. Through the one level feature layer data fusion and the two-level decision layer data fusion, the three-level fusion method obtained assessment results for the overall health status of the EMU through monitoring data of the bogie and its auxiliary system, and the traction system, and verified the accuracy and effectiveness of the method. It has definite application value. However, this method also has certain problems that need to be improved. For example, the contribution of components to the overall evaluation is subjective; as the breadth and depth of EMU monitoring increases, the data quality increases, it is necessary to consider the influencing factors more comprehensively. How to solve the above problems will become the next research direction.

REFERENCES

[1] ISO-13374-1:2003. Condition Monitoring and Diagnostics of Machines—Data Processing, Communication and Presentation.

[2] D.L. Hall, J. Llinas, An introduction to multisensor data fusion, Proc. IEEE 85 (1) (1997) 6–23.

[3] R.C. Luo, M.G. Kay, Multisensor integration and fusion in intelligent systems, IEEE Trans. Syst. Man Cybern. 19 (5) (1989) 901–931.

[4] K. Yaohong, Data Fusion Theory and Applications, Xidian University Press, Xi'an, 1997.

[5] L. Yu, C. Hai, Z. Qingjiang, Z. Kairui, Research on the method of UAV system health assessment, J. Syst. Eng. Electron. 33 (3) (2011) 562–567.

[6] L. Zhenzhu, X. Zhicheng, X. Weihong, et al., Transformer health status assessment method considering the remaining life of insulation, Power Autom. Equip. 36 (8) (2016) 137–142.

[7] Q. Wenxi, H. Kaoli, L. Guangyao, et al., Improved DSmT health status assessment based on uncertainty and importance, J. Aerospace Pow. 32 (1) (2017) 96–104.

[8] K. Shouqiang, W. Yujing, C. Lili, L. Changyuan, Z. Jianbiao, Method for evaluating health status of rolling bearings based on CFOA-MKHSVM, Chin. J. Sci. Instrum. 37 (9) (2016) 2029–2035.

[9] H. Manuo, F. Shangchun, Z. Dezhi, X. Weibiao, Research progress of multisensor data fusion, Sens. Microsyst. 29 (3) (2010) 5–8.

[10] Y. Xin, F. Hua, Coal and gas outburst prediction based on case-based reasoning and data fusion, J. Southeast Univ. 41 (Suppl) (2011) 59–63.

[11] Q. Shuang, W. Yi, Research on data fusion algorithms in wireless sensor networks, J. Wuhan Univ. Technol. 30 (7) (2008) 119–122.

[12] X. Bojie, Z. Huaping, Coal mine safety status assessment based on multi-sensor data fusion, Comput. Modernization 222 (2) (2014) 138–141.

[13] Hanwen Cheng, Chen Liang, Lenan Wu, Decision-level fusion multi-system modulation recognition based on weighted voting, J. Syst. Eng. Electron. 32 (2) (2010) 342–345.

[14] D. Xupu, et al., Mathematical Modeling Method and Practice, National Defense Industry Press, Beijing, 2006.

Index

Note: Page numbers followed by *f* indicate figures and *t* indicate tables.

F

G

H

Kolibri 360 GmbH
Ferdinand-Jühlke-Straße 7
99095 Erfurt, Deutschland
produktsicherheit@kolibri360.de